# Distance Geometry and Conformational Calculations

# CHEMOMETRICS RESEARCH STUDIES SERIES

*Series Editor:* **Dr. D. Bawden**
*Pfizer Central Research, Sandwich, Kent, England*

# Distance Geometry and Conformational Calculations

**Dr. G.M. Crippen**

*Assistant Professor,*
*Department of Chemistry,*
*Texas A & M University, U.S.A.*

**RESEARCH STUDIES PRESS**
A DIVISION OF JOHN WILEY & SONS LTD.
Chichester · New York · Brisbane · Toronto

RESEARCH STUDIES PRESS

*Editorial Office:*
8 Willian Way, Letchworth, Herts SG6 2HG, England

Copyright © 1981, by John Wiley & Sons Ltd.

*British Library Cataloguing in Publication Data:*

Crippen, G.M.
    Distance geometry and conformational calculations.
    —(Chemometrics research studies series)
    1. Mathematical physics
    2. Stereochemistry
    I. Title    II. Series
    530.1'5    QC20.8    80-42044

    ISBN 0 471 27991 9

Printed in Great Britain

# TABLE OF CONTENTS

## ACKNOWLEDGEMENTS

This work originated during my stay in the Department of Pharmaceutical Chemistry at the University of California, San Francisco. The development of the ideas owes a great deal to the enthusiasm and stimulation of I. D. Kuntz, my close collaborator in these studies. Other important contributors are T. F. Havel, M. L. Connolly, P. A. Kollman, R. Langridge, and E. C. Jorgensen. This manuscript and some of the illustrations were prepared using the UCSF Computer Graphics Laboratory facilities (supported by NIH Biotechnology Resources Grant RR-1081, Principal Investigator: Robert Langridge).

## INTRODUCTION

Distance geometry refers to the study of geometric problems with an emphasis on the distances between points. Recently this has been successfully applied to a variety of conformational calculations which are much more difficult by conventional methods. In order to explain the special properties of conformational calculations by distance geometry, it will be necessary to review the usual techniques.

Suppose we are interested in the conformation of a molecule consisting of $n$ atoms. Clearly one can always specify their positions in space with respect to some fixed Cartesian coordinate system, using $3n$ variables. The conformational properties, however, are dependent only on the *relative* positions of the atoms, so the problem is reduced by 3 translational and 3 rotational degrees of freedom to $3n-6$ variables. Although these are a necessary and sufficient set of variables for exactly expressing the conformation, many of the degrees of freedom are nearly constant, such as bond stretching and bond angle bending. These can be eliminated simply by holding all bond lengths and angles at their energetically preferred values and allowing only rotation about appropriate bonds. As a simple example, ethane's two carbon and six hydrogen atoms would require 24 Cartesian coordinates or 1 dihedral angle, namely the rotation about the C-C bond. The specification of conformation is more succinct, and calculations involving a smaller number of variables are generally easier. Thus the policy has been to reduce the number of variables as far as possible without removing degrees of freedom that are of physical interest.

## 2 Introduction

In the case of small molecules, employing dihedral angles is a fine idea, but for larger ones, numerical problems arise. Considering Cartesian coordinates to be the primary variables, a bond length, $d_{ij}$, is calculated as the magnitude of the bond vector between atoms $i$ and $j$, $\mathbf{v}_{ij}$, which is the first order difference of the coordinates, $\mathbf{c}_i$.

$$d_{ij} = \|\mathbf{v}_{ij}\| = \|\mathbf{c}_i - \mathbf{c}_j\|$$

Since the bond angle, $\theta_{ijk}$, defined by the three atoms $i$, $j$, and $k$, involves in turn a comparison of the two bond vectors $\mathbf{v}_{ij}$ and $\mathbf{v}_{jk}$, it is a second order difference quantity with even higher relative error.

$$\theta_{ijk} = \arccos(\hat{\mathbf{v}}_{ij} \cdot \hat{\mathbf{v}}_{jk}),$$

where $\hat{\mathbf{v}}$ is the normalized vector. Similarly, the dihedral angle $\tau_{ijkl}$ defined by the four points $i$, $j$, $k$, and $l$ is a comparison of the normals of the $ijk$ and $jkl$ planes, and these normals are themselves calculated by comparisons of bond vectors, making the dihedral angle a third order difference quantity.

$$\tau_{ijkl} = \arccos[(\hat{\mathbf{v}}_{ij} \times \hat{\mathbf{v}}_{jk}) \cdot (\hat{\mathbf{v}}_{jk} \times \hat{\mathbf{v}}_{kl})]$$

Clearly calculating dihedral angles from coordinates propagates whatever errors there might be, much as numerical calculation of third derivatives would. Working in the other direction, error propagation is equally bad, since calculating coordinates from dihedral angles is analogous to extrapolating the values of a function from initial conditions and its third derivative. For large molecules the instability is particularly noticeable, not only because a long series of dihedral angles specify the path that a chain takes through space, but also because of the "lever arm" effect. If rotation about some bond by an angle $\Delta\tau$ moves an atom which lies at a distance $r$ from the bond axis, then that atom will move the distance $r\sin\Delta\tau$. When $r$ is large enough, that can be a substantial motion, even for small $\Delta\tau$. In general, there is increasing numerical instability upon going to more compact representations of conformation. On the other hand, the set of all interatomic distances are only first order differences of the coordinates and also contain a great deal of redundancy for a large number of points, since there are only $3n-6$ degrees of freedom but $n(n-1)/2$ interpoint distances.

Another important but frequently neglected feature of choosing parameters is the effect on other functions of the system. Ordinarily one wants to calculate the intramolecular energy of a molecule as a function of the conformational variables. However, the number of local minima (and hence the difficulty of finding energetically favorable conformations) depends on the variables used. In empirical energy calculations the energy is customarily calculated by

# 3   Introduction

$$E = \sum_{i<j}^{n} f(d_{ij})$$

for some function $f$ of the pairwise interatomic distances. In my experience the energy is a rather complicated function of the dihedral angles, having numerous local minima, whereas the surface as a function of Cartesian coordinates is much smoother, and it is simpler still when viewed as a function of the distances directly. A well documented case of this general phenomenon can be found in the study of Griewank et al. (1979), who show that the energy for positioning a rigid molecule in a crystal has more saddle points and minima when treated as a function of the 3 Euler angles (the minimum number of variables) than when quaternions (4 variables) are used.

To be fair, the reader should be warned at the outset of the disadvantages of the distance geometry approach. The number of variables is so large that examining and comprehending the conformational space is rather difficult without computer algorithms to help. One can draw a graph of the intramolecular energy of ethane as a function of its single dihedral angle, but exploring all 28 interatomic distances is not so easy. Since the number of distances goes as the square of the number of atoms, computer memory requirements become a problem for large molecules. The better optimization algorithms require work spaces on the order of the square of the number of variables, so energy minimization with respect to a large number of variables can be difficult to arrange. The greatest difficulty, however, is that not every choice of interatomic distances corresponds to a realizable arrangement of the atoms in space. It is this embedding problem that will concern us for most of the rest of the book.

The general problem that distance geometry addresses is to calculate atomic coordinates for a molecule given constraints in terms of upper and lower bounds on the interatomic distances. The result will be a random sampling of satisfactory conformers, rather than the single "best" one, since ordinarily whole regions of conformation space are permitted. Usually the constraints are derived from experimental results, and may be of low accuracy, but preferably affect many of the distances. The technique has been slanted toward the case where no particular atom deserves greater emphasis than another, and the given constraints apply to many different parts of the molecule. The chapters that follow explain in considerable detail how to carry out conformational calculations by distance geometry, but initially a qualitative outline of the method will help keep it in perspective:

1.  Use the given upper bounds to deduce some sort of upper bound on every distance.

2.   Similarly the given lower bounds imply some sort of lower bound for every distance.

3.   Choose random trial distances between their respective upper and lower bounds for every interatomic distance.

4.   Calculate center-of-mass, principle-axes initial coordinates which reproduce the trial distances as closely as possible.

5.   Refine the coordinates until they completely satisfy the original constraints.

Each step and illustrative applications will be covered in the subsequent chapters.

## 2.1 Basic Linear Algebra

Before going into detail about the problem of embedding a distance matrix in three dimensional space, some may appreciate a quick review of the fundamental notation and results of linear algebra, since matrices and vectors play a central role in what follows. No attempt is made at giving proofs or thoroughly developing the subject, as that is better done in the many text books on the subject.

A *scalar* is simply an ordinary number, and will always be denoted by a lower case italic symbol, such as $a$ or $\lambda$. A *vector* **v** of dimension $n$ consists of $n$ scalar elements in a specific order: $\mathbf{v} = (v_1, v_2, \ldots, v_n)$. Vectors will always be denoted by a lower case bold face letter. The geometric interpretation of a vector is that of an arrow pointing from the origin of a coordinate system in $n$ dimensions to a point whose coordinates are the elements of the vector. The *norm* or length of **v** is defined by

$$\|\mathbf{v}\| \equiv \left( \sum_{i=1}^{n} v_i^2 \right)^{1/2} .$$

Addition of two vectors of the same dimension is defined elementwise:

$$\mathbf{u} + \mathbf{v} \equiv (u_1 + v_1, \ldots, u_n + v_n) .$$

Geometrically, the distance from the point indicated by **u** to the point at **v** is calculated by $\|\mathbf{v} - \mathbf{u}\|$. A vector may be multiplied by a scalar:

$$a\mathbf{v} \equiv (av_1, \ldots, av_n) .$$

In this way a vector may always be *normalized* to unit length (denoted by ^ ):

$$\hat{\mathbf{v}} \equiv \mathbf{v}/\|\mathbf{v}\| \; .$$

There is a *scalar product* (or *dot* product) between two vectors of the same order:

$$\mathbf{u} \cdot \mathbf{v} \equiv \sum_{i=1}^{n} u_i v_i$$

but the *vector* (or *cross*) *product* is defined only for $n=3$:

$$\mathbf{u} \times \mathbf{v} \equiv \begin{bmatrix} u_2 v_3 - u_3 v_2 \\ u_3 v_1 - u_1 v_3 \\ u_1 v_2 - u_2 v_1 \end{bmatrix} \; .$$

The right side of the equation has been written as a *column vector* where the elements run from top to bottom instead of from left to right. Unless otherwise stated, all vectors are to be thought of as column, rather than row, vectors. Clearly $\mathbf{v} \cdot \mathbf{u} = \mathbf{u} \cdot \mathbf{v}$, but $\mathbf{v} \times \mathbf{u} \neq \mathbf{u} \times \mathbf{v}$ in general. If $\mathbf{v} \cdot \mathbf{u} = 0$, then $\mathbf{v}$ and $\mathbf{u}$ are *orthogonal*.

A *matrix* is an $n \times m$ array of scalar elements, and is denoted by an upper case bold face symbol:

$$\mathbf{A} = \begin{bmatrix} a_{1,1} & a_{1,2} & \cdot & \cdot & a_{1,m} \\ a_{2,1} & a_{2,2} & \cdot & \cdot & a_{2,m} \\ \cdot & \cdot & & & \cdot \\ \cdot & \cdot & & & \cdot \\ a_{n,1} & a_{n,2} & \cdot & \cdot & a_{n,m} \end{bmatrix}$$

The first index refers to the row, the second to the column. $\mathbf{B} = \mathbf{A}^T$, the *transpose* of $\mathbf{A}$, is an $m \times n$ matrix with the indices interchanged. That is, $b_{ji} = a_{ij}$. The transpose of a column vector is a row vector, and vice versa. A *square matrix* of order $n$ is an $n \times n$ matrix. Such a matrix $\mathbf{C}$ may be in addition *symmetric* if $\mathbf{C} = \mathbf{C}^T$. A *real* matrix has no complex elements. Matrices may be added or subtracted in the elementwise fashion already indicated above for vectors, if the numbers of rows and columns are respectively equal. For the $l \times m$ matrix $\mathbf{A}$ and the $m \times n$ matrix $\mathbf{B}$, *matrix multiplication*, $\mathbf{AB} = \mathbf{C}$, is defined as

$$c_{ij} = \sum_{k=1}^{m} a_{ik} b_{kj},$$

where $\mathbf{C}$ is an $l \times n$ matrix. A column vector is simply the special case of an $n \times 1$ matrix, so a matrix times a column vector = a column vector, and similarly a row vector times a matrix = a row vector.

It may occur that $\mathbf{Av} = \lambda \mathbf{v}$ (or $\mathbf{v}^T \mathbf{A} = \lambda \mathbf{v}^T$). Then the scalar $\lambda$ is an *eigenvalue* of $\mathbf{A}$, and $\mathbf{v}$ is the corresponding (right) *eigenvector* of $\mathbf{A}$ (or

$\mathbf{v}^T$ is the left eigenvector). If $\mathbf{A}$ is real, symmetric, and of order $n$, then the left and right eigenvectors are the same, and there are $n$ real, but not necessarily distinct, eigenvalues. The corresponding eigenvectors are automatically orthogonal for distinct eigenvalues, and those for equal eigenvalues may always be chosen to be orthogonal. If $\mathbf{v}$ is an eigenvector, then so is $s\mathbf{v}$ for any scalar, $s$, which means that eigenvectors can always be normalized. Hence a real symmetric matrix of order $n$ has an orthonormal set of $n$ eigenvectors. The *spectral expansion* of $\mathbf{A}$ is then

$$\mathbf{A} = \sum_{i=1}^{n} \lambda_i \mathbf{v}_i \mathbf{v}_i^T$$

The *determinant*, $|\mathbf{A}|$, of the square matrix $\mathbf{A}$ of order $n$ is a polynomial consisting of terms each containing $n$ factors of elements of $\mathbf{A}$. It is defined recursively as follows: For $n=2$,

$$|\mathbf{A}| \equiv a_{1,1}a_{2,2} - a_{1,2}a_{2,1} .$$

Let $\mathbf{A}]_{ij}$ be the $(n-1)\times(n-1)$ matrix obtained by deleting row $i$ and column $j$ from $\mathbf{A}$. Then

$$|\mathbf{A}| \equiv \sum_{i=1}^{n} (-1)^{i-1} |\mathbf{A}]_{i1}| .$$

It can be shown that $|\mathbf{A}| = \prod_{i=1}^{n} \lambda_i$. When the determinant is zero, there is a zero eigenvalue, and the matrix is in some sense "redundant". Deleting a row and a column from such a *singular* matrix may produce a *minor* of $\mathbf{A}$ which has a non-zero determinant. The order of the largest submatrix with non-zero determinant is the *rank* of $\mathbf{A}$, and is equal to the number of its non-zero eigenvalues.

## 2.2 Blumenthal's Theorem

If one is to propose a distance matrix $\mathbf{D}$ of distances $d_{ij}$ between $n$ points (atoms or groups of atoms, for our purposes), then $\mathbf{D}$ must have certain properties regardless of how distances are calculated and how many spatial dimensions are allowed:

(1)  $\mathbf{D}$ is a symmetric $n \times n$ matrix: $d_{ij} = d_{ji}$.

(2)  Diagonal elements are all zero: $d_{ij} = 0$.

(3)  All off-diagonal elements are strictly greater than zero: $d_{ij} > 0$, $i \neq j$ (otherwise some points $i$ and $j$ would be identical and a smaller order $\mathbf{D}$ would be appropriate).

Since $\mathbf{D}$ is symmetric, it is sufficient to refer to only the upper triangle in all that follows. It is possible to propose a distance matrix **D** satisfying all the conditions given above, and yet not have it correspond to a realizable conformation. As a simple example, observe that there is no *two*-dimensional arrangement of four points corresponding to

$$\mathbf{D} = \begin{vmatrix} 0 & 1 & 1 & 1 \\ 1 & 0 & 1 & 1 \\ 1 & 1 & 0 & 1 \\ 1 & 1 & 1 & 0 \end{vmatrix}.$$

The matrix is that of the corners of a tetrahedron, requiring three dimensions. Analogous problems arise when trying to embed an $n$-tuple of points in ordinary three-dimensional Euclidean space, $E_3$. The necessary and sufficient conditions for embedding $n$ points in $E_r$ for any given $r$, are given in a theorem due to Blumenthal (1970). We cite it here specialized to three dimensions, with appropriate changes in nomenclature.

Theorem (Blumenthal, 1970) *A necessary and sufficient condition that a semimetric $(n+1)$-tuple may be irreducibly congruently embeddable in $E_3$ $(3 \leqslant n)$ is that an ordering of the points exists so that (i) sgn $\Delta(1,...,k+1) = (-1)^{k+1}$ for all $k=1, \ldots, m \leqslant 3$, and (ii) for any other points $u$ and $v$, $m+1 \leqslant u, v \leqslant n+1$, then $\Delta(1, \ldots, m+1, u) = \Delta(1, \ldots, m+1, v) = \Delta(1, \ldots, m+1, u, v) = 0$. Here $\Delta(1, \ldots, k)$, etc. is the Cayley-Menger determinant of the distances $d_{ij}$ between the k points*

$$\Delta(1, \ldots, k) = \begin{vmatrix} d_{11}^2 & d_{12}^2 & \cdots & d_{1k}^2 & 1 \\ d_{21}^2 & d_{22}^2 & & & \cdot \\ \cdot & \cdot & & \cdot & \cdot \\ \cdot & \cdot & & \cdot & \cdot \\ \cdot & \cdot & & \cdot & \cdot \\ d_{k1}^2 & d_{k2}^2 & \cdots & d_{kk}^2 & 1 \\ 1 & 1 & \cdots & 1 & 0 \end{vmatrix}.$$

Remarks.   The theorem is expressed so as to allow for a possibly coplanar ($m=2$) or colinear ($m=1$) set of points. Clearly, any set of points lying in a lower dimensional Euclidean space can also be embedded in a higher dimensional one. Otherwise ($m=3$), the theorem proceeds by first choosing some quartet of the points that are not coplanar, and then numbers them 1, 2, 3, and 4. The embedding conditions for the other points are then made in reference to this basis quartet.

Corollary 1 (Crippen, 1977).    *Part (i) is true for k=1 as long as* $d_{12}^2 > 0$, *as it must.*

$$\Delta(1,2) = \begin{vmatrix} 0 & d_{12}^2 & 1 \\ d_{21}^2 & 0 & 1 \\ 1 & 1 & 0 \end{vmatrix} = 2d_{12}^2 \ .$$

Corollary 2 (Crippen, 1977).    *Part (i) for k=2 and m⩾2 is equivalent to saying the n+1 points are not all collinear and the distances between the first three points satisfy the triangle inequality*

$$d_{12} + d_{23} \geqslant d_{13}$$

*for any permutation of the subscripts 1, 2, and 3.*

*Proof.* Squaring the inequality twice, one obtains

$$d_{12}^4 + d_{23}^4 + d_{13}^4 - 2d_{12}^2 d_{13}^2 - 2d_{23}^2 d_{13}^2 - 2d_{12}^2 d_{23}^2 \leqslant 0$$

but this expression is exactly the result of evaluating $\Delta(1,2,3)$. Equality holds in both inequalities only if the three points are colinear.

Corollary 3 (Crippen, 1977). *Part (i) for k=m=3 is equivalent to the n+1 points being not all coplanar and the distances between the first four points satisfying what might be called the "quadrangle inequality". Denote the various squared distances by lower case unsubscripted letters thus:*

$$\Delta(1,2,3,4) = \begin{vmatrix} 0 & a & c & d & 1 \\ a & 0 & b & e & 1 \\ c & b & 0 & f & 1 \\ d & e & f & 0 & 1 \\ 1 & 1 & 1 & 1 & 0 \end{vmatrix} \ .$$

Note that $d = d_{14}^2$. *Then referring to Fig. 2.1, the quadrangle inequality says that d must lie between* $d_{min}$ *(when the points are in the planar cis configuration (as illustrated) with respect to rotation about the 2-3 bond, and* $d_{max}$ *(when in the planar trans configuration), assuming all other distances a, b, c, e, and f, to be fixed.*

*Proof.* Evaluating the determinant, one obtains

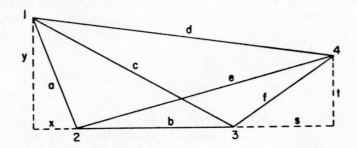

Fig. 2.1 Four numbered points with mutual squared distances denoted a, b, c, d, e, and f. The arrangement shown is planar cis with respect to rotation about the 2-3 bond. See the proof of Corollary 3 for the use of squared distances x, y, s, and t.

$$\Delta(1,2,3,4) = 2[-a^2f-b^2d-c^2e-d^2b-e^2c-f^2a \tag{2.1}$$

$$+bde+bce+cef+cde+aef+ace$$

$$+abf+abd+acf+bdf+bcd+adf$$

$$-bef-ade-abc-cdf] \ .$$

Now referring to Fig. 2.1, if we take all distances but $d^{1/2}$ to be fixed, then the quadrangle inequality says that $d^{1/2}$ can vary between its lower bound $d_{min}^{1/2}$ in the planar *cis* conformation, as illustrated, and its upper bound $d_{max}^{1/2}$ in the planar *trans* conformation by rotating about the 2-3 "bond". We now calculate $d_{max}$ and $d_{min}$ using Fig. 2.1 and the pythagorean theorem

$$x+y = a \tag{2.2}$$

(all distances are squared) and

$$c = y+(b^{1/2}+w_c x^{1/2})^2 \tag{2.3}$$

where

$$w_c = \begin{cases} +1 & \text{for } c>a+b \\ -1 & \text{for } c<a+b \end{cases} \tag{2.4}$$

Eliminating $y$ and solving for $x$,

$$x^{1/2} = \frac{c-a-b}{2w_c b^{1/2}} \tag{2.5}$$

and similarly

$$s^{1/2} = \frac{e-f-b}{2w_e b^{1/2}} .$$

(2.6)

Hence, from eqs. 2.5 and 2.6

$$d_{min} = (b^{1/2}+w_c x^{1/2}+w_e s^{1/2})^2+(y^{1/2}-t^{1/2})^2$$

(2.7)

$$= (1/4b)(c-a+e-f)^2+(y^{1/2}-t^{1/2})^2 .$$

Similarly, we obtain

$$d_{max} = (1/4b)(c-a+e-f)^2+(y^{1/2}+t^{1/2})^2.$$

(2.8)

Combining eqs. 2.3 and 2.5 we find

$$y = c-\frac{(c+b-a)^2}{4b}$$

(2.9)

and similarly

$$t = e-\frac{(e+b-f)^2}{4b} .$$

(2.10)

Substituting eqs. 2.9 and 2.10 into eq. 2.7 yields after simplification

$$d_{min} = \frac{1}{2b}\{[ce-cf-ae+af+bc+ba-b^2+be+bf]$$

(2.11)

$$-[4cb-(c+b-a)^2]^{1/2}[4eb-(e+b-f)^{1/2}]\} ,$$

and similarly eqs. 2.8, 2.9, and 2.10 give an expression for $d_{max}$ that differs only in the sign between the square brackets. From the quadrangle inequality we know that

$$-2b(d-d_{min})(d-d_{max}) \geqslant 0,$$

(2.12)

but substituting in eq. 2.11 and the equivalent expression for $d_{max}$ yields the [right hand side of eq. 2.1] $\geqslant 0$ after much expansion and simplification.

Corollary 4. (Crippen, 1977)    *Part (ii) of the theorem is satisfied if the bordered matrix* $D(1, \ldots, n+1)$ *of squared distances corresponding to* $\Delta(1, \ldots, n+1)$ *has rank* $m+2$. *In particular, for* $n+1 > 4$ *points not all coplanar located in* $E_3$, *the rank of* $D$ *must be 5.*

*Proof.*    The rank of a matrix is the order of its largest nonzero minor. If the rank of $D$ is $m+2$, then the points can be numbered so that for the first $m+1$ of them, $\Delta(1, \ldots, m+1) \neq 0$, and $\Delta(1, \ldots, m+1, u) = \Delta(1, \ldots, m+1, v) = \Delta(1, \ldots, m+1, u, v) = 0$, since all minors of order higher than $m+2$ must be zero. (Note that the border adds an extra row and column.)

Note that the converse of corollary 4 is also true (Aitken, 1956).

## 2.3 Whole Matrix Embedding

The emphasis in Blumenthal's theorem is on setting up a basis of 4 non-coplanar points and then positioning the remaining points relative to them. This is fine when dealing with a full set of exact interpoint distances, but in practice the distances are subject to some uncertainty, and there is usually no reason to single out some four points as special and particularly accurate. Therefore the policy has been to devise methods of embedding which make use of all the distances at once.

Suppose we have $n$ points and the corresponding $n \times n$ distance matrix, $\mathbf{D}$. Let $\mathbf{C}$ be the derived bordered matrix of squared distances, as shown in theorem 2.1. Since $\mathbf{C}$ is a real, symmetric matrix of order $n+1$, there are $n+1$ (real) eigenvalues, which we may number in descending order of absolute value:

$$|\lambda_1| \geqslant |\lambda_2| \geqslant \cdots \geqslant |\lambda_{n+1}| .$$

Let $\mathbf{E}$ be the $(n+1) \times (n+1)$ matrix whose columns are the corresponding eigenvectors. Then

$$c_{ij} = \sum_{k=1}^{n+1} e_{ik} \lambda_k e_{jk}$$

for $i,j = 1, \ldots, n+1$. Now the rank 5 requirement for embeddability given in corollary 4 in the previous section would be satisfied if the summation over eigenvalues and eigenvectors were simply truncated at 5, since this would correspond to having $|\lambda_5| > \lambda_6 = \cdots = \lambda_{n+1} = 0$. Thus we calculate the five eigenvalues largest in absolute value (by numerical methods to be explained in chapter 3) and form the modified $\mathbf{C}$ matrix:

$$c'_{ij} = \sum_{k=1}^{5} e_{ik} \lambda_k e_{jk} .$$

This algorithm (Crippen, 1978) produces $\mathbf{C}'$, the rank 5 matrix closest to $\mathbf{C}$ in the spectral sense, which qualitatively means that the larger elements tend to be preserved, so that the gross features of the desired conformation are approximately achieved (often at the expense of the small-scale structure, such as bond lengths). In addition, the diagonal elements of $\mathbf{C}'$ may not be zero, and the border elements may deviate from unity. Observe that there are the correct number of degrees of freedom: [5 variable eigenvalues] + [$5(n+1)$ variable components of the five eigenvectors] - [$n+1$ constraints of zero diagonal elements] - [$n$ constraints of unity border elements] - [15 constraints from the orthonormality of the five eigenvectors] = $3n-6$ net degrees of freedom.  Note that $\mathbf{C}'$ is automatically symmetric.

Deviations from the desired values of the diagonal and border elements of $\mathbf{C}'$ are numerically difficult to correct.  In my experience the matrix has been "well-conditioned" with respect to perturbations of the eigenvalues but ill-conditioned with respect to eigenvectors.  Nevertheless, one can use $\mathbf{C}'$ straight away to calculate coordinates for the points, $\mathbf{v}_i = (x_i, y_i, z_i)$, $i = 1, \ldots, n$, taking $c'_{ij}$ to be the squared distance, $d_{ij}^2$ for all $i$, $j$.

$$\mathbf{v}_1 = (0,0,0)$$

$$\mathbf{v}_2 = (d_{12}, 0, 0)$$

$$\mathbf{v}_3 = \left( \frac{d_{13}^2 - d_{23}^2 + d_{12}^2}{2d_{12}} \, , \, [d_{13}^2 - x_3^2]^{1/2} \, , \, 0 \right)$$

$$\mathbf{v}_4 = \left( \frac{d_{14}^2 - d_{24}^2 + d_{12}^2}{2d_{12}} \, , \, \frac{d_{24}^2 - d_{34}^2 + d_{13}^2 - d_{12}^2 + 2x_4(d_{12} - x_3)}{2y_3} \, , \right.$$

$$\left. [d_{14}^2 - x_4^2 - y_4^2]^{1/2} \right)$$

$$\mathbf{v}_i = \left( \frac{d_{1i}^2 - d_{2i}^2 + d_{12}^2}{2d_{12}} \, , \right.$$

$$\frac{d_{2i}^2 - d_{3i}^2 - d_{12}^2 + d_{13}^2 + 2x_i(d_{12} - x_3)}{2y_3} \, ,$$

$$\left. \pm [d_{1i}^2 - x_i^2 - y_i^2]^{1/2} \right) \text{ for } i = 5, \ldots, n$$

(2.13)

where the sign of $z_i$ is chosen so that $|d_{4i} - \|\mathbf{v}_i - \mathbf{v}_4\| \, |$ is minimal.  Note that one could choose $z_4$ to be negative, thus reversing the handedness of the whole resulting figure.  The distance matrix of a collection of points is unique only up to a rigid translation, rotation, or mirror inversion.  The inherent numerical instability of eq. 2.13 is readily apparent, for example when $d_{13} \approx d_{23} >> d_{12}$, and small errors in the $d$'s result in large errors in $x_3$.  Furthermore, although $\mathbf{C}'$ is an overall approximation to $\mathbf{C}$, the calculation of coordinates make explicit use of only the first four rows of $\mathbf{C}'$.

The preferred method (Crippen & Havel, 1978) for embedding a proposed distance matrix $\mathbf{D}$ in $E_3$ employs the "metric matrix", $\mathbf{G}$, to increase numerical stability and avoid emphasizing any particular distances. Let some point, "$o$", be chosen as the origin of a coordinate system. Then if $\mathbf{u}_i$ is the vector from the origin to atom $i$ in the molecule,

$$g_{ij} \equiv \mathbf{u}_i \cdot \mathbf{u}_j \ . \tag{2.14}$$

Clearly $\mathbf{G}$ is closely related to the coordinates, but in fact it can easily be calculated from the distance matrix using the law of cosines:

$$g_{ij} = \frac{1}{2}(d_{io}^2 + d_{jo}^2 - d_{ij}^2) \ .$$

Let $\mathbf{W}$ be the 3 columned matrix containing the 3 (mutually orthogonal) eigenvectors of the real symmetric $\mathbf{G}$ corresponding to the three eigenvalues, $\lambda_1$, $\lambda_2$, and $\lambda_3$, of largest absolute value.

**Theorem 2.2**    *Then the three coordinates of point $i$, $v_{ij}$, for $j$=1, 2, 3 are given by*

$$v_{ij} = \lambda_j^{1/2} w_{ij} \ . \tag{2.15}$$

*Proof.*    Observe that

$$g_{ij} = \sum_{k=1}^{n} w_{ik} w_{kj} \lambda_k$$

when *all* eigenvalues and eigenvectors have been calculated. But if we knew the coordinates of the points in $n$-dimensional space, then

$$g_{ij} = \sum_{k=1}^{n} v_{ik} v_{jk}$$

by the definition of the metric matrix. Equating the corresponding terms in the two sums yields eqn. 2.15.

In order for eq. 2.15 to give real coordinates, it is necessary for the first three eigenvalues of $\mathbf{G}$ to be non-negative. Experience has shown that as long as the proposed distances do not violate the triangle inequality for any triplet of points, this is almost always the case (Braun et al., 1980). We can prove this is true only in the very simple situation of three points.

**Theorem 2.3**    *Label three points 0, 1, and 2. If the metric matrix with point 0 as the origin has a negative eigenvalue, then there is a violation of the triangle inequality among the distances.*

*Proof.*    Solving the secular equation,

$$\begin{vmatrix} g_{11}-\lambda & g_{12} \\ g_{12} & g_{22}-\lambda \end{vmatrix} = 0$$

$$\lambda^2 + (-g_{11}-g_{22})\lambda + (g_{11}g_{22}-g_{12}^2) = 0$$

$$\lambda = \frac{1}{2}[g_{11}+g_{22}\pm\sqrt{(g_{11}-g_{22})^2+4g_{12}^2}\,]$$

(note that $\lambda$ must always be real because the discriminant must be non-negative) we find that $\lambda < 0$ means that

$$\sqrt{(g_{11}-g_{22})^2+4g_{12}^2} > g_{11}+g_{22}$$

which yields upon simplification

$$g_{12}^2 > g_{11}g_{22} \ .$$

Converting to distances,

$$\frac{1}{2}(d_{10}^2+d_{20}^2-d_{12}^2) > d_{10}d_{20}$$

and rearranging

$$(d_{10}-d_{20})^2 > d_{12}^2$$

which implies one of

$$d_{10} > d_{12}+d_{20}$$

$$d_{20} > d_{10}+d_{12}$$

$$d_{12} > d_{20}+d_{10} \ .$$

Note that for $n$ points, the metric matrix has order only $n-1$, since one of the points must be chosen as the origin. The point of origin may easily be changed without reference to the original distances. Let **G** be the metric matrix with respect to point "o" and **H** with respect to point "x". Then from the definition of **G** in eq. 2.14, $g_{ii}=d_{io}^2$ and hence

$$d_{ij}^2 = g_{ii}+g_{jj}-2g_{ij} \tag{2.16}$$

Substituting into the equivalent expression for $h_{ij}$, we find

$$h_{ij} = g_{ij}+g_{xx}-g_{ix}-g_{jx} \ . \tag{2.17}$$

Choosing one of the original $n$ points as the origin for **G** can be an advantage if it is important to emphasize the distances from that point. As with the previous method, embedding via **G** weights the greater distances throughout the structure. Suppose, however, that the short distances to the origin point were known to high accuracy, and the long distances were to be deemphasized. Any transformation of the form

$$h_{ii} = f_i(g_{ii}) \tag{2.18}$$

$$h_{ij} = \frac{\sqrt{h_{ii}h_{jj}}\,g_{ij}}{\sqrt{g_{ii}g_{jj}}} \quad \text{for } i \neq j$$

yields a weighted metric matrix $\mathbf{H}$, which preserves the angles between the vectors from the origin to the points while varying their lengths. If $\mathbf{H}$ is reduced to rank 3 in the usual way, and the resultant matrix is transformed back (assuming $f$ has an inverse) then the result also has rank 3. The difference is that the large entries in $\mathbf{H}$ were emphasized in embedding, not the large elements of $\mathbf{G}$. The only restrictions on the functions $f_i$ are that they be continuous and positive valued over positive arguments, and that they have an inverse. As indicated, one need not even choose the same function for each point $i$. For example, to weight short distances from the origin, one might choose $h_{ii} = w_i g_{ii}^2$ and hence $g'_{ii} = (h_{ii}/w_i)^{1/2}$, for some positive weights, $w_i$.

The more usual situation is that there is no one point whose distances should *a priori* be stressed in the final structure. Then it is best to take the center of mass of the molecule as the origin, rather than one of the atoms. Distances from the center of mass point "0" can be calculated directly from the interpoint distances without reference to coordinates (Crippen & Havel, 1978):

Theorem 2.4 (Havel)

$$d_{i0}^2 = \frac{1}{n}\sum_{j=1}^{n}d_{ij}^2 - \frac{1}{n^2}\sum_{j=2}^{n}\sum_{k=1}^{j-1}d_{jk}^2 \;. \tag{2.19}$$

*Proof.*   Let $\mathbf{r}_{kl}$ denote the vector from point $l$ to point $k$; $n =$ total number of points; and 0 denotes the center of mass point. From the definition of the center of mass of an array of points, each of unit mass:

$$\sum_{j=1}^{n}\mathbf{r}_{j0} = 0 = \sum_{j=1}^{n}(\mathbf{r}_{10}+\mathbf{r}_{1j})$$

so that

$$\mathbf{r}_{10} = -n^{-1}\sum_{j=2}^{n}\mathbf{r}_{1j}$$

and

$$d_{10}^2 = \mathbf{r}_{10}\cdot\mathbf{r}_{10} = n^{-2}\sum_{j=2}^{n}\sum_{k=2}^{n}\mathbf{r}_{1j}\cdot\mathbf{r}_{1k} \;.$$

By the law of cosines,

$$d_{10}^2 = (2n^2)^{-1} \sum_{j=2}^{n} \sum_{k=2}^{n} (d_{1j}^2 + d_{1k}^2 - d_{jk}^2)$$

$$= (2n^2)^{-1} \left[ 2(n-1) \sum_{j=2}^{n} d_{1j}^2 - 2 \sum_{j=2}^{n} \sum_{k=j+1}^{n} d_{jk}^2 \right]$$

$$= \frac{n-1}{n^2} \sum_{j=2}^{n} d_{j1}^2 - n^{-2} \sum_{j=2}^{n} \sum_{k=j+1}^{n} d_{jk}^2$$

$$= n^{-1} \sum_{j=1}^{n} d_{j1}^2 - n^{-2} \sum_{j=1}^{n} \sum_{k=j+1}^{n} d_{jk}^2 \quad .$$

Since the labelling of the points is arbitrary, then we have the general formula, eq. 2.19 . It is possible with erroneous distances to calculate negative distances to the center of mass. Although eq. 2.19 holds for any number of dimensions of Euclidean space (the proof makes reference only to vectors to the points and their scalar products, not to their dimensionality), there are choices of distances which are not embeddable in *any* number of dimensions. Even satisfying the triangle inequality for all triplets of points is insufficient to guarantee positive distances to the center of mass, as is shown by the following counterexample (Connolly, 1977):

$$\mathbf{D} = \begin{bmatrix} 0 & 3 & 3 & 3 \\ 3 & 0 & 6 & 5 \\ 3 & 6 & 0 & 5 \\ 3 & 5 & 5 & 0 \end{bmatrix}$$

Clearly the triangle inequality is obeyed throughout $\mathbf{D}$, but applying eq 2.19 gives $d_{10} = -5/16$.


## 2.4 Embedding with Constraints

Suppose we have some sort of upper bound, $u_{ij}$, (not necessarily the least upper bound) and lower bound, $l_{ij}$, (not necessarily the greatest lower bound) on each interpoint distance, $d_{ij}$. The experimental and theoretical sources for this distance information will be discussed in chapter 4. One would like to calculate the coordinates of points in $E_3$ that have mutual distances obeying $l_{ij} \leqslant d_{ij} \leqslant u_{ij}$ for all $i, j$. It may be that some of the upper bounds are stronger than others, and that simply applying the triangle inequality to all triplets of points,

$$u_{ij} \leqslant u_{ik} + u_{kj} \tag{2.20}$$

may allow improvement of some of the bounds. Thus, if the left side of eq. 2.20 is greater than the right, then $u_{ij}$ may be decreased until equality holds. Applying this procedure successively to all triplets of points, $i$, $j$, and $k$, monotonically lowers or leaves unchanged the elements of the upper bound matrix, **U**. In a few complete passes, no further changes can be made. Geometrically, if one pictures the points as beads interconnected with strings whose lengths correspond to the $u_{ij}$, then the final triangle inequality consistent upper bound between $i$ and $j$ is the distance when the network of strings is pulled taut, holding beads $i$ and $j$. Once **U** is determined, the analogous inequality on the lower bounds,

$$l_{ij} \geqslant l_{ik} - u_{jk} \tag{2.21}$$

can be used to raise some of the $l_{ij}$. A trial distance matrix can then be built up by, say, randomly choosing each $d_{ij}$ independently with uniform distribution from the corresponding interval $[l_{ij}, u_{ij}]$. The difficulty is that even if eqs. 2.20 and 2.21 are satisfied by the bounds, the resulting trial distances need not even satisfy the triangle inequality, much less the other necessary embedding conditions. Calculated distances to the center of mass may be negative, and some of the first three eigenvalues of **G** may be negative. The basic problem is that the interpoint distances cannot be completely independently chosen, since there are more distances than degrees of freedom and the embedding conditions constitute their interrelations. A full knowledge of the correlations between distances is tantamount to a complete understanding of the embedding problem, which is very difficult for a large number of points. However, some of the strongest correlations can be calculated on the triangle level of sophistication.

Suppose for points $i$, $j$, and $k$, we are given the mutual upper and lower distance bounds, and we want to calculate the correlation coefficients among $d_{ij}$, $d_{ik}$ and $d_{jk}$ according to a method by Connolly (Crippen et al., 1980). Each triple of distances, $(d_{ij}, d_{ik}, d_{jk})$ can be identified with a point $(x,y,z)$ in the positive octant of three-dimensional space. Upper and lower bounds on the three distances limit the permitted volume to a rectangular parallelopiped, henceforth to be referred to as a "cube", although its sides need not be of equal length. The triangle inequalities further limit the permitted volume to those points lying on the correct side of the three triangle inequality planes $x+y=z$, $x+z=y$, and $y+z=x$.

There are three possibilities for the set of points allowed by these constraints: (1) no points allowed -- the bounds are inconsistent, (2) the entire cube is permitted -- the distances are uncorrelated, or (3) an irregular convex polyhedron remains -- the distances are correlated. We now show how to calculate the correlation coefficients from integrals over the convex polyhedron.

## 19  Mathematics of Embedding

Let $X$ represent the random variable associated with the probability distribution of the distance $d_{ij}$, and similarly let $Y$ and $Z$ represent the $d_{ik}$ and $d_{jk}$ distributions, respectively. The correlation coefficients can be calculated in terms of the variances and covariances, e.g.:

$$corr(X,Y) = \frac{covar(X,Y)}{\sqrt{var(X)\ var(Y)}}$$

The variances and covariances are in turn defined in terms of expectation values:

$$var(X) = E(X^2) - E(X)^2$$

$$covar(X,Y) = E(XY) - E(X)E(Y)$$

where the expectation values are integrals of probability density functions (p.d.f.) over the permitted volume. Since we assume the distances to be distributed uniformly between the lower and upper bounds, our p.d.f. is the reciprocal of the permitted volume, $V$. The expectation values are then simply the first and second moments of the convex polyhedron, normalized by dividing by the volume.

$$E(X) = \int \frac{1}{V} x\, dV$$

$$E(XY) = \int \frac{1}{V} xy\, dV$$

$$E(X^2) = \int \frac{1}{V} x^2\, dV$$

Since the shape of the domain of integration (the convex polyhedron) depends upon relations between the lower and upper bounds on the distances, there is no simple expression for the values of these integrals. An integral over the polyhedron can be decomposed, however, into an integral over the cube minus three integrals over the corner pieces cut off the cube by the triangle inequality planes, there being no overlap between the pieces cut off by different planes (Fig. 2.2). Each corner piece can itself be decomposed as an alternating sum of tetrahedra, one for each vertex of the cube which lies on the forbidden side of the triangle inequality plane. Such a vertex is called forbidden, because its coordinates fail one of the triangle inequalities. Each of these tetrahedra is defined by a forbidden vertex and by three mutually perpendicular edges issuing from that vertex and parallel to the coordinate axes which extend to meet the plane. The sign in front of each term of the alternating sum is plus or minus according to the sign of the vertex, as defined below. This alternating sum is an application of the inclusion-exclusion principle (Hall, 1967). The various moments of these tetrahedra can be calculated by considering them as scaled and translated versions of a standard tetrahedron located at the

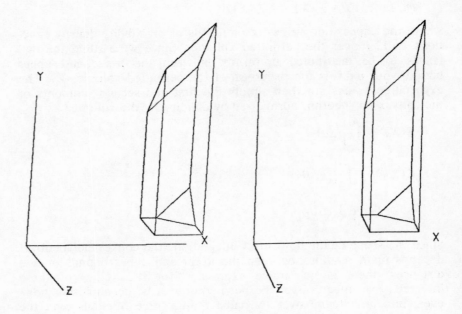

Fig. 2.2.   Convex polyhedron for distance bounds: $6 \leqslant d_{ij} \leqslant 9$, $3 \leqslant d_{ik} \leqslant 20$, $4 \leqslant d_{jk} \leqslant 8$. The triangle inequality limits $d_{ik}$ (Y) substantially, $d_{ij}$ (X) somewhat, and $d_{jk}$ (Z) not at all.

origin with coordinates: $(0,0,0)$, $(1,0,0)$, $(0,1,0)$, and $(0,0,1)$. Similarly, the moments of the cube can be calculated from the moments of a standard unit cube at the origin. If we label the coordinates of a given forbidden vertex $v$ of the cube by $v_0, v_1, v_2$; the signed lengths of the three perpendicular edges meeting $v$ by $e_0, e_1, e_2$; the (positive) Jacobian determinant of the scaling transformation by $J$; and the volume, first and second moments of the standard tetrahedron by $\bar{V}, \bar{F}, \bar{S}_{ij}$; then the volume, first and second moments of the tetrahedron, $V^v$, $F_i^v$, $S_{ij}^v$, are given by:

$$V^v = \bar{V}J , \quad J = |e_0 e_1 e_2|$$

$$F_i^v = (v_i + e_i \bar{F})J , \quad i=0,1,2$$

$$S_{ij}^v = \{v_i v_j + (e_i v_j + e_j v_i)\bar{F} + e_i e_j \bar{S}_{ij}\}J , \quad i,j=0,1,2$$

The moments of the cube are given by the same formulas, except with different values of $\bar{V}$, $\bar{F}$, and $\bar{S}_{ij}$. The values for the standard tetrahedron and cube are given in the table below.

| Shape | $V$ | $F$ | $S_{ij}, i=j$ | $S_{ij}, i \neq j$ |
|-------|-----|-----|---------------|--------------------|
| Tetrahedron | 1/6 | 1/24 | 1/60 | 1/120 |
| Cube | 1 | 1/2 | 1/3 | 1/4 |

We can now write down formulas for the various expectation values, which we relabel for conciseness as $E_0 = E(X)$, $E_{01} = E(XY)$, $E_{00} = E(X^2)$, etc.

$$V = V^{cube} - \sum_{\substack{forbidden \\ vertices\ v}} sgn(v) V^v$$

$$E_i = \frac{1}{V}\left\{ F_i^{cube} - \sum_{\substack{forbidden \\ vertices\ v}} sgn(v) F_i^v \right\}$$

$$E_{ij} = \frac{1}{V}\left\{ S_{ij}^{cube} - \sum_{\substack{forbidden \\ vertices\ v}} sgn(v) S_{ij}^v \right\}$$

$Sgn(v)$ is defined to be $+1$ if the number of the coordinates of the vertex $v$ which are lower bounds is even, and $-1$ if it is odd. A vertex $v$ is forbidden if $v_0 > v_1 + v_2$, $v_1 > v_2 + v_0$, or $v_2 > v_0 + v_1$.

## 3.1 Eigenvalues

A central step in the embedding calculations outlined in section 2.3 was determining the three (or five) eigenvalues of largest absolute value for a real, symmetric matrix, $\mathbf{C}$, of order $n$, where $n$ may be large. There are a number of ways this can be done, but the situation is particularly fortunate since we do not need *all* the eigenvalues and their eigenvectors, but only the largest ones. The algorithm we have used is Faddeev & Faddeeva's (1963) "exhaustion" procedure. Starting with an order $n$ vector, $\mathbf{y}$, of random components, repeatedly premultiply by $\mathbf{C}$. Then the largest magnitude eigenvalue, $\lambda_1$, and its eigenvector $\mathbf{v}_1$ may be approximated by iterating $k$ times, for $k=1,2,...$

$$\frac{(\mathbf{C}^k \mathbf{y}) \cdot (\mathbf{C}^k \mathbf{y})}{(\mathbf{C}^{k-1} \mathbf{y}) \cdot (\mathbf{C}^k \mathbf{y})} \rightarrow \lambda_1, \quad as \ k \rightarrow \infty$$

and $\mathbf{C}^k \mathbf{y} \rightarrow \mathbf{v}_1$ (unnormalized). Then the next smaller eigenvalue, $\lambda_2$ is found similarly by using $\mathbf{C} - \lambda_1 \mathbf{v}_1 \mathbf{v}_1^T$ in the place of $\mathbf{C}$, and so on. Convergence of the $\lambda$'s to six significant figures is regularly achieved well before $k=100$, even when two eigenvalues are close in magnitude (which generally slows convergence in this method). The computational effort involved in the matrix-vector multiplication goes up as $n^2$, but since only the first three (or five) eigenvalues are sought, the cost otherwise depends linearly on $n$.

## 3.2 Triangle Inequality

The ranges of distances allowed in a conformational problem may often be greatly restricted by first exhaustively applying the triangle inequality (eq. 2.20) to the upper bounds, and its analog (eq. 2.21) to the lower bounds. As explained in section 2.4, the exhaustive application soon converges, but since all triplets of points must be considered, the computing time is proportional to $n^3$. In spite of the computing expense, it is well worth while, particularly when little *a priori* information is available about some distances, in which case the initial upper bounds must be very large ($\infty$), and the lower bounds must be zero.

Choosing random trial distances, $d_{ij}$, from the corresponding range, $[l_{ij}, u_{ij}]$, is done with a suitably scaled and translated "pseudo-random" number generator with uniform distribution over the interval. Such routines are readily available in most Fortran libraries, but one may be easily constructed (Hammersley & Handscomb, 1964) using the linear congruential method. Given a "seed" number, $a_1$, the pseudo-random sequence of numbers, $a_2$, $a_3$, ... is generated, where $0 \leqslant a_i \leqslant 1$ for i=2,3,...

$$a_{i+1} = \frac{(e + d[ba_i]) \bmod c}{b}$$

where $[x]$ = largest integer $\leqslant x$. For a 32 bit computer, suitable choices of the constants are: $b=1073741823$, $c=1073741824$, $d=282475253$, and $e=226908346$.

The trial distances may either be chosen at random independently, or with some correlation. The method of calculating precise triangle-inequality-level correlation coefficients, as outlined in section 2.4, is too lengthy to be applied to all triplets of points in a large problem. There are however, two main situations when the coefficients are significantly different from zero: (1) if $u_{jk} \leqslant 0.2 u_{ik}$ then the correlation, $c_{ij,ik} \approx 0.9$ ; (2) if $l_{jk} \geqslant 0.9 u_{ik}$, then $c_{ij,ik} \approx -0.5$ . In an effort to apply all correlations from all triplets uniformly at once, we have used the following approximation: let $a_{ij}$ be the random number for the $i,j$ distance, chosen uniformly from the interval [0,1]. Then choose trial distances by

$$d_{ij} = l_{ij} + (u_{ij} - l_{ij}) \left[ \frac{a_{ij} + \sum_{case\ 1} c_{ij,ik} a_{ik} + \sum_{case\ 2} c_{ij,ik} (a_{ik} - 1)}{1 + \sum_{case\ 1} 0.9 + \sum_{case\ 2} 0.5} \right] .$$

The result is that the trial distances tend to be distributed more heavily near the center of the allowed range, much as the sum of uniformly distributed random numbers has a Gaussian distribution. The disadvantage that the extremes of conformation space tend not to be sampled is generally outweighed by the greatly improved trial distances. Calculated distances to the center of mass are always positive in my experience, negative eigenvalues of **G** occur much less frequently, and

## 25   Numerical Methods

the initial coordinates calculated from the rank 3 **G** have many fewer violations of the distance bounds.

### 3.3 Optimization

The initial coordinates, $c_i$, calculated according to eq. 2.15 by applying the rank 3 condition to the metric matrix, are a good approximation to the trial distances, but there is no guarantee that they satisfy the original distance constraints. The violations may be eased by minimizing the scalar function $f(v_1, \ldots, v_n)$ with respect to the $3n$ Cartesian coordinates.

$$f(v_1, \ldots, v_n) = \sum_{j>i} \begin{cases} (d_{cij}^2 - u_{ij}^2)^2, & d_{cij} > u_{ij} \\ 0, & l_{ij} \leqslant d_{cij} \leqslant u_{ij} \\ (d_{cij}^2 - l_{ij}^2)^2, & d_{cij} < l_{ij} \end{cases} \qquad (3.1)$$

where $d_{cij} = \|c_i - c_j\|$ . As written, $f$ tends to stress the violation of large distances. Of course any sort of positive weights may multiply the terms without altering the facts that $f=0$ is its global minimum (perhaps unattainable if the bounds are not satisfied by any three-dimensional structure), $f=0$ implies all constraints are satisfied, and $f>0$ otherwise. For example, dividing each term in eq. 3.1 by the corresponding $u_{ij}^4$ would weight each distance equally so that molecular bond lengths and angles are more accurately reproduced.

There are many numerical optimization algorithms available, but the better ones often require computer memory space proportional to the square of the number of variables. For a molecule represented as $n$ points (atoms or groups of atoms), there are $3n$ variables in the optimization stage, where $n$ may be more than 100. We have obtained the best results by beginning the minimization of $f$ with the method of steepest descents, which quickly reduces the value of $f$ and its gradient, $\nabla f$, but does not converge on the minimum rapidly. When $\|\nabla f\|^2$ drops below some cutoff, such as $10^{5\ 6}$, it is better to switch to the more rapidly converging conjugate gradient algorithm (Fletcher & Reeves, 1964). Both of these minimizers require storage only proportional to the first power of the number of variables.

For molecular conformational calculations, the chirality of the molecule is often important. However, the distance matrix for a particular conformation is unchanged by a mirror inversion of the entire molecule. Thus equivalent conformations of stereoisomers yield the same distance matrix. If there is more than one chiral center in the molecule, then technically the distance matrices of diastereomers *are* distinguishable, just as they have different physical properties with respect to an achiral environment. In other words, distance geometry cannot distinguish the DL isomer from the LD one, but it can tell the difference between DL and DD. In practice, the situation is even somewhat worse, because one generally calculates coordinates starting from the distance bounds, U and L, where there is often a substantial range allowed for most distances. Then the difference between diastereomers is a relatively subtle correlation of distances among the substituents of the two chiral centers, all well within the bounds. The simplest treatment is to add another "chirality violation" penalty term to eq. 3.1, as follows.

Stereoisomerism arises whenever four distinguishable points in a molecule are constrained to lie in relatively fixed positions with respect to one another, and these positions are not all in one plane. Suppose these four points are the substituents of an asymmetric carbon atom, and they are labeled 1 through 4 in ascending rank according to the Cahn-Ingold-Prelog system (Cahn et al., 1966). Denote by $\mathbf{v}_i$ the position vector of point $i$. Then the scalar, $f_{ch}$, defined by

$$f_{ch} \equiv (\mathbf{v}_1 - \mathbf{v}_4) \cdot [(\mathbf{v}_2 - \mathbf{v}_4) \times (\mathbf{v}_3 - \mathbf{v}_4)] \tag{3.2}$$

is 6× the *signed* volume of the tetrahedron formed by the four points. If the asymmetric center has the R configuration, $f_{ch}$ will be negative; $f_{ch}$ has the same absolute value but a positive sign in the S configuration. Correct chirality of the final structure is forced by including in eq. 3.1 the extra terms

$$\sum_{chiral\ centers} (f_{ch} - f_{ch}^*)^2 \tag{3.3}$$

where $f_{ch}^*$ is corresponding desired value (magnitude and sign) of $f_{ch}$. Although correcting the chirality of the initial coordinates requires only a local puckering, sometimes other terms in eq. 3.1 temporarily inhibit the movement in the chiral center. This can be avoided by either weighting the chiral constraints more heavily or by first minimizing their contribution to the penalty function alone.

## 3.4 Summary Algorithm

Suppose we wish to calculate one or more conformations of a molecule given distance constraints for some (but not all, in general) pairs of points (atoms). The procedure is as follows:

1.  Let the initial values of the $u_{ij}$ be set to some large number, for all unknown upper bounds. Otherwise the given values are used, of course. Reduce those upper bounds violating the triangle inequality, eq. 2.20, as explained in section 2.4 until no further alterations can be made.

2.   Starting with the given lower bounds and otherwise a default of $l_{ij}=0$, exhaustively increase those lower bounds violating eq. 2.21 until no further alterations can be made.

3.   Choose correlated trial distances according to the method of section 3.2 . In general different random numbers will result in different trial distances and hence final coordinates, so it is this stage that allows one to explore the range of allowed conformations at random.

4.   Calculate the distance to center of mass for each point, using eq. 2.19 . Then convert the $n \times n$ matrix of distances to the corresponding $n \times n$ metric matrix, $\mathbf{G}$, with center of mass as the origin, according to the equation following eq. 2.14 .

5.   Find the three eigenvalues of $\mathbf{G}$ having largest absolute value and their corresponding eigenvectors using the exhaustion algorithm of section 3.1, and then calculate the initial coordinates of the $n$ points by eq. 2.15 .

6.   The initial coordinates do not in general still satisfy the original constraints, so one must now minimize the penalty function, eq. 3.1, with respect to coordinates. The upper and lower bounds used are the triangle "smoothed" ones of steps 1 and 2 above. Any chirality penalty terms (eq. 3.3) must be included.

The final coordinates at the end of step 6 represent a randomly chosen conformation of the molecule which satisfies all the input constraints. For most applications, the computer program that carries out these calculations may be considered as a "black box". The lengthy discussions so far have been intended for the investigator whose needs require some tailoring of the usual algorithm. The next chapter is concerned with how to bridge the gap between experimental data and the distance constraints the computer calculations must begin with.

## 4.1 Encoding Data

Distance geometry conformational calculations require input in the form of upper and lower bounds on some of the interatomic distances. However, the available experimental evidence may be expressed in terms of angles and shapes, for example. The purpose of this section is to help the investigator make the conversions into distance bound form.

*Bond lengths* are usually known *a priori* from x-ray crystallography. If the length of the bond between atoms $i$ and $j$ is known to be $b \pm e$, then clearly $u_{ij} = b + e$ and $l_{ij} = b - e$. If the error $e \approx 0$, the calculations work just as well setting $u_{ij} = l_{ij} = b$.

Expressing the *ijk bond angle* requires fixed $ij$ and $jk$ bond lengths. Then by the law of cosines,

$$d_{ik}^2 = d_{ij}^2 + d_{jk}^2 - 2d_{ij}d_{jk}\cos\theta_{ijk}$$

so that setting $\theta_{ijk}$ to the appropriate value fixes $d_{ik}$.

In a similar fashion, the *dihedral angle,* $\tau_{hijk}$ (i.e., for the chain of atoms $h-i-j-k$, the torsional angle for rotation about the $ij$ bond) is a restriction on $d_{hk}$, once all the other distances among the four atoms are fixed; or equivalently, when the $hi$, $ij$, and $jk$ bond lengths, and the $hij$ and $ijk$ bond angles are fixed. If the *cis* conformation defines $\tau_{hijk} = 0$, then

$$d_{hk}^2 = (d_{ij} - d_{hi}\cos\theta_{hij} - d_{jk}\cos\theta_{ijk})^2 +$$

$$(d_{hi}\sin\theta_{hij} - d_{jk}\sin\theta_{ijk}\cos\tau_{hijk})^2 +$$

$$(d_{jk}\sin\theta_{ijk}\sin\tau_{hijk})^2$$

Note that $d_{hk}$ is unchanged when $-\tau_{hijk}$ is substituted for $\tau_{hijk}$. If the information input to the conformational problem really distinguishes between these two cases, then the distance geometry algorithm must employ a chiral center consisting of atoms $h-k$.

As already discussed in section 3.3, a *chiral center* is treated in a rather *ad hoc* fashion as an extra penalty term during the optimization step. Whenever there are four atoms whose mutual distances are nearly fixed, and the input to the problem warrants fixing their chirality, then a penalty term (eq. 3.3) must be added to eq. 3.1, where $f_{ch}^*$ is calculated from coordinates for the four atoms using eq. 3.2 . The precise magnitude of $f_{ch}^*$ depends on the bond lengths and angles among the atoms, so in order to be consistent with the other distance constraints, the correct value must be calculated from coordinates which may be either taken from x-ray crystal data or calculated using eq. 2.13 . Since only four atoms are required, the asymmetric carbon itself or any one of its substituents is superfluous. Also note that each atom in a molecule is treated as unique in these calculations, so that for example methane could have chiral constraints even though *chemically* the hydrogens are indistinguishable.

In general, a *rigid group* of atoms contained within the molecule, such as a phenyl ring or even an $\alpha$-helical segment in a protein, can be specified by setting interatomic upper and lower bounds equal to the distances found in model structures. The distance geometry algorithm in principle needs only barely enough constraints to completely specify the rigid group, but in practice convergence is more rapid if *all* distances among atoms in the group are fixed. As soon as there are more than three non-coplanar atoms in the group, the possibility of chiral constraints arises, in addition to the pure distance constraints.

*Crosslinking* constraints are particularly easy to incorporate into the calculation. For example, if a protein is represented at low resolution as only one point per residue, located at the $C^\alpha$, then a disulfide bridge between residues $i$ and $j$ implies that $u_{ij}=6.5$ and $l_{ij}=6.0$ . Knowing that a long but rather flexible crosslinking reagent can join two parts of a molecule or molecular complex is often helpful in that the upper bound may be substantially reduced, but there would be still a large range between the upper and lower bounds.

Distance geometry is ordinarily able to deal only with geometric constraints between *labelled* points in the molecule. Thus crosslinking is important information when it is known which two atoms are linked, but when it is only known that there is some pair which can be thus connected, it is difficult to incorporate the information into the calculations. A useful exception is when the information applies to *all* pairs of atoms. For example, if one knows from hydrodynamic experiments or small angle x-ray diffraction what the *diameter* of the molecule must be, then this value can be simply used as the default upper bound for all pairs of atoms.

In the event there is some information available about distances of points to the center of mass of the molecule, the easiest approach is to add on an extra point to the problem which explicitly represents the center of mass. This trick works rigorously only when there are sufficiently many molecule points whose distances to the center of mass are constrained so that the extra point indeed finally lies near the center of mass calculated from the coordinates of all the atoms. As an extreme example, suppose we only knew the distance from center of mass to one point, out of a total of 100 molecule points. Then point 101, representing the center of mass, could well lie completely outside the molecule. In such a situation, eq. 2.19 must be used to interrelate the trial distances between molecule points, and the extra center of mass point is no longer necessary.

When more is known about the overall *shape* of the molecule than its spherical diameter, *outrigger points* are a convenient way of incorporating the information. These are extra points beyond those necessary to represent the molecule itself, which have fixed, large mutual distances. The shape of the molecule is expressed by confining all molecule points to lie within the intersection of appropriately chosen spherical shells centered at the outriggers. Although the outriggers could in principle be carried through all stages of the distance geometry algorithm (section 3.4) it has been my experience that convergence in the final penalty function minimization step can be substantially improved by fixing the outrigger coordinates so as to exactly satisfy their desired mutual distances and then minimize with respect to only the molecule point coordinates. As an example (Crippen, 1979a) the coat protein monomer of tobacco mosaic virus is known to occupy in the intact virus a trapezoidal slab 25 thick along the virus particle's cylinder axis, 22° wide, running from 20 to 90 radially out from the axis. The TMV protein itself was represented by 79 points (one for every two amino acid residues) and four outrigger points were chosen, #80-83. These were positioned as shown in fig. 4.1 to define the required shape. The distances among the outriggers,

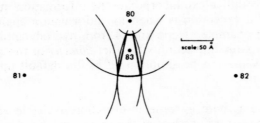

Figure 4.1   Plan view of the TMV protein wedge (trapezoid in heavy lines) showing the placement of outrigger points 80—83 to scale.   Arcs indicate the upper and lower bound distances allowed between points of the polypeptide chain and the various outrigger points.   Point 83 is 200   out of plane.

$$
\mathbf{D}_{outrig} = \begin{array}{c|cccc}
 & (80) & (81) & (82) & (83) \\
(80) & 0 & 188.272 & 188.272 & 205.760 \\
(81) & & 0 & 332.513 & 263.137 \\
(82) & & & 0 & 263.137 \\
(83) & & & & 0
\end{array}
$$

place #80 at the cylinder axis so that $u_{80,i}=90$ for $i=1,...,79$ determines the bottom of the wedge shaped allowed region of fig. 4.1, and $l_{80,i}=20$ determines the top.   Outrigger points 81 and 82 were located to the sides by means of appropriately chosen 80—81, 80—82, and 81—82 distances so that $u_{81,i},u_{82,i}=183.429$ and $l_{81,i},l_{82,i}=149.084$ confined the protein to the experimentally determined 22° wedge shape.   Point 83 was then placed out of the plane of the previous three outriggers to confine the protein to the desired 25   thickness of wedge by means of $u_{83,i}=225$ and $l_{83,i}=200$ .

A certain degree of "*unlabelling*" of molecule points can be tolerated by modifying the penalty function, eq. 3.1 .   Consider two disjoint sets of molecule points, $S_I=\{i_1, \ldots , i_k\}$ and $S_J=\{j_1, \ldots , j_l\}$, having the constraint that for at least one $i \epsilon S_I$ and $j \epsilon S_J$, $l_{IJ} \leqslant d_{ij} \leqslant u_{IJ}$. Then let the contribution from this constraint to the total penalty function, eq. 3.1, be given by

$$
\min_{\substack{i \epsilon S_I \\ j \epsilon S_J}}
\begin{cases}
(d_{ij}^2 - u_{IJ}^2)^2, & d_{ij} > u_{IJ} \\
0, & l_{IJ} \leqslant d_{ij} \leqslant u_{IJ} \\
(d_{ij}^2 - l_{IJ}^2)^2, & d_{ij} < l_{IJ}
\end{cases}
$$

For example (Kuntz and Crippen, 1980), some of the proteins in the 30S ribosome subunit are known to be near each other by chemical crosslinking and neutron diffraction evidence, but it is also apparent that the individual proteins are extended, rather than spherical. Therefore they were represented as three points each, and distance information between the protein indicated by points in $S_I$ and the protein of $S_J$ was introduced in the optimization of penalty function stage by a term as given above.

Arbitrarily complicated overall *shape constraints* can be built into the penalty function if one is willing to devise a coordinate system in terms of three outrigger points, and a "shape function", $S(x,y,z)$. Then for all molecule points $i$, one adds to eq. 3.1 the terms

$$\sum_i [R_i^2 - S(x,y,z)^2]^2$$

where $R_i$ is the distance between point $i$ and the $z$-axis, and the coordinates of point $i$ are $x,y,z$. For very complicated figures, $S(x,y,z)$ may no longer be single valued, so it becomes necessary to define more than one (parallel) $z$-axis and assign the least penalty for point $i$ tested against each $z$-axis and its corresponding shape function.

## 4.2 Examples of Applications

In the previous section I have already alluded to a study on the effectiveness of the experimental studies on tobacco mosaic virus protein toward determining its structure. As of 1975 (before x-ray crystal structures became available) there was a great deal of conformational evidence on this cylindrical virus: x-ray diffraction studies on oriented gels of the virus (yielding radial distances for some residues), electron microscopy, hydrogen exchange, immunochemistry, various kinds of spectroscopy, and determination of accessibility to attacking reagents. The protein itself was given a very coarse representation as a single connected chain of 79 points, one for every two amino acid residues. Chain connectivity determined 78 upper bounds, $u_{i,i+1}$, and the self avoiding property (space filling) determined a default lower bound on the residues. Outrigger point 80, located on the cylinder axis, provided a reference for specifying 16 radial distances of labelled residues. All four outriggers were used to set the overall shape requirement, as already discussed above. Five random conformers fulfilling all constraints were generated and compared according to the root-mean-square deviation of their distance matrices, $\Delta$ (which conveniently is independent of translation and rigid rotation of one structure with respect to the other). If the coordinates of conformation $A$ are designated by vectors $\mathbf{a}_i$, $i=1,...,n$; and those of conformation $B$ are given by $\mathbf{b}_i$, $i=1,...n$; then $\Delta(A,B)$ is calculated by the formula

$$\Delta(A,B) = \left[\frac{\sum_{i=1}^{n-1}\sum_{j=i+1}^{n}(\|\mathbf{a}_i-\mathbf{a}_j\| - \|\mathbf{b}_i-\mathbf{b}_j\|)^2}{0.5n(n-1)}\right]^{1/2} \tag{4.1}$$

The significance of this work with regard to distance geometry technique is that (i) minor inconsistencies in the input from different experimental techniques were detected as residual errors in the final coordinate optimization and as triangle inequality violations; (ii) otherwise all the constraints could be accommodated without difficulties in the calculations even though the data came from such diverse sources; and (iii) among the five generated structures there were some similar conformational features, but in overall rms deviation, they were grossly different in spite of the numerous experimental constraints. This last outcome is generally a desirable feature of distance geometry calculations, enabling the investigator to avoid overestimating the effectiveness of the evidence toward limiting the range of possible conformations.

In a paper by Havel *et al.* (1979), various sorts of experimental and theoretical constraints were simulated in advance of actually having methods to honestly produce them. Pancreatic trypsin inhibitor (58 residues) and carp myogen (108 residues) were represented at the single point per residue level, placing the point at the $C^\alpha$. The chain connectivity and space filling constraints are known in advance, but other sorts of information could be abstracted from the known x-ray crystal structures of these two proteins. For example, a theory which could correctly predict which residues are in contact was simulated by setting the upper bound to 10   for all residue pairs which were within 10   in the crystal structure. The surprising outcome was that such *qualitative* information could produce *quantitatively* correct results ( $\Delta$(generated, native) $\approx 1$ ). Correct secondary structure prediction was simulated by taking $\alpha$-helices and $\beta$-strands as rigid units from the crystal structure. Center of mass information was encoded for all residues using the extra center of mass point method. Massive crosslinking results (66 distances) were simulated in the obvious way. The above sorts of constraints and combinations of them were used for altogether 20 different trials, each resulting in 10 structures, without excessive computer time, and with average bound violations of less than 0.1 .

As an example of a more detailed conformational calculation, NMR evidence on the conformation of the glycopeptide antibiotic bleomycin was used to generate possible structures (Crippen et al., 1980). A minor part of the molecule was not included in the calculations in any way, but the rest was represented in great detail, using one point for each of the 75 non-hydrogen atoms. The default upper bounds was 250 , and the default lower bound 2.8  (a conservative estimate of the Van der Waals diameter of a $CH_2$ group). All bond lengths and bond angles were fixed at values derived from appropriate crystal structures, which also ensured planar carbonyl groups, tetrahedral $sp^3$ carbon atoms, etc. The NMR evidence indicated that the gulose and mannose residues contained in bleomycin have invariant conformations, so all distances among atoms in these groups were fixed at values obtained from appropriate crystal structures. The planarity of two conjugated rings was further enforced by including all distances among the atoms of each ring in the list of constraints. Three peptide bonds were fixed to be planar *trans* by setting appropriate 1—4 distances. The 19 asymmetric carbon atoms in the molecule were given their correct configuration by including the appropriate chiral terms in the penalty function. Six proposed ligand groups to a metal ion were expressed in the ordinary way, and octahedral ligation was encouraged by an appropriate choice of lower bounds among the ligand atoms. As it turned out, this constraint was not very strong. NMR experiments indicated that a certain (assigned) proton was close to the imidazole ring of bleomycin, which corresponds to small upper bounds on all the distances from the carbon attached to the proton to all the imidazole atoms. In addition to requiring the imidazole and pyrimidine ring nitrogens to coordinate to the metal, we insisted that they do so only in the direction of the nitrogen lone pair orbitals. This was done by fixing the distance between the metal and the atom on the opposite side of the ring from the nitrogen to the value obtained when metal, N, and opposite side of ring are colinear. The dihedral angles for rotation about two other bonds were restricted in accordance with the observed proton-proton coupling. All told there were approximately 300 constraints on the conformation. The major difficulty encountered was that sometimes the trial coordinates were near the correct full set of desired distances for one of the sugar rings, but in the mirror-image chirality. Then the distance constraints were fully satisfied and resisted the efforts of the chiral constraints to invert the entire ring. This could be avoided by weighting the chiral constraints more heavily. Furthermore, satisfying all the detailed constraints caused rather lengthy optimizations in the last stage of the calculation, so that structures required an average of 47 seconds apiece on a CDC 7600 computer. Precise bond lengths and angles were obtained only after optimizing until the error function, eq. 3.1, had a very small value. Presumably appropriate weights on the error terms would have helped. Careful comparison of the few generated structures revealed a surprising variety of conformation, considering how difficult it is to build a space filling physical model of the molecule which incorporates all the above constraints. Since only 5 structures were calculated, it was difficult to speak with certainty about invariant conformational features, as they might have been coincidental.

The 30S subunit of the ribosome study (Kuntz and Crippen, 1980) involved some novel adaptations, as already discussed above. The experimental data used came from either immuno-electron microscopy or neutron scattering. The electron microscopy information came in the form of coordinates in an established reference frame, which was used for the shape function and positioning of some of the proteins. Three points per protein allowed us to represent each protein as appropriately extended, but required the semi-labelled penalty terms explained earlier. The neutron scattering data gives not just upper and lower bounds between parts of two proteins, but the distribution of lengths of vectors connecting some part of one with some part of another. To a first approximation, the extremes of this distribution were used to determine the $u_{IJ}$'s and $l_{IJ}$'s used for the partially labelled constraints. In addition, the center of the distribution was taken as the desired distance between the central point of the one protein triple and the central point of the other protein triple. In order to do this, one of the three points representing a protein had to carry the extra label of "central". Of course it is a considerable simplification to represent an extended protein as only three points and to use only three parts of an entire distribution of distances, but it is an illustration of how to deal with such kinds of data within the distance geometry framework. Once again, small inconsistencies in the input data were easily detected and were resolved by essentially assigning somewhat larger error bars to the data. The optimization stage appeared to be extremely difficult, due mostly to the complicated shape function. Unlike most distance geometry applications, local minima were frequently encountered and required small perturbations of the coordinates in order to continue optimizing.

## 5.1 Physical Model

One of the most important applications of distance geometry has been in modelling the binding of small ligand molecules to sites on macromolecules, such as proteins. One of the major problems of medicinal chemistry is to deduce structural details and binding energetics of a site given the experimentally determined free energies of binding for a series of analogues. There are a number of methods presently used for this sort of quantitative structure-activity relation (QSAR) study, but the distance geometry approach is unexcelled in its treatment of the spatial aspects of the problem.

I make the following assumptions (Crippen, 1979b): (i) binding is observed to occur on a single site of a pure receptor protein; (ii) each ligand has a well-determined chemical structure and stereochemistry but may be flexible due to rotation about single bonds; (iii) no chemical modification of the ligands occurs during the binding experiment, although the conformation of the ligand may change upon binding to fit the binding site; (iv) the free energy of such a conformational change is small compared to the free energy of binding; (v) the experimentally determined free energy of binding is given and is approximately the sum of the "interaction energies" for all "contacts" between parts of the ligand molecule and parts of the receptor site; and (vi) the site itself may be slightly flexible, although no major conformational changes are permitted, and the energetic cost of any deformation is negligible.

---

Material in this chapter is reprinted in part with permission from *J. Med. Chem.* *22*, 988 (1979), copyright 1979, American Chemical Society; and *J. Med. Chem. 23*, 599 (1980), copyright 1980, American Chemical Society.

The ligand is represented in the usual way as a collection of points in space. For example in fig. 5.1 we see the decomposition of $m,m$-$(CH_3)_2C_6H_3OCH_2COCH_3$ into five points. A very precise description of a molecule in these terms would be to position a point at the nucleus of every atom; however, such precision may not be necessary, and it does lead to more expensive calculations later on. At the other extreme, one may choose to approximate whole groups of atoms, for example an entire benzene ring, as a single point located perhaps at its center of mass. This is an adequate description if the binding of the ligand molecules is rather nonspecific, such as having only a large hemispherical pocket where a benzene ring may be positioned equally well in a variety of orientations. On the other hand, suppose the hydrophobic pocket is shaped like a narrow slot so that the ring may now be inserted in only one orientation. Then it would be necessary to represent the benzene ring as at least three points, perhaps positioned at $C(1)$, $C(3)$, and $C(5)$. In general, larger numbers of points will be required to describe a ligand molecule when the fit in the site is more specific and intricate. With the present algorithms, one discovers that the site is rather specific only by noting that simple descriptions of the ligand molecules fail to account for the binding data.

For some choice of representing the ligands as points corresponding to atoms or groups of atoms, we describe the entire ensemble of conformations the ligand may attain by the upper and lower bounds on the interpoint distances. Since both the **U** and **L** matrices are symmetric with zeros along the diagonal, we can compactly represent both in a single bound matrix, **B**, where the lower triangle contains the lower bounds and the upper triangle the upper bounds. Table 5.1 shows **B** for $m,m$-dimethylphenoxyacetone. Representing even conformationally variable drug molecules by simply an upper and lower bound distance matrix is an approximation, since sometimes there

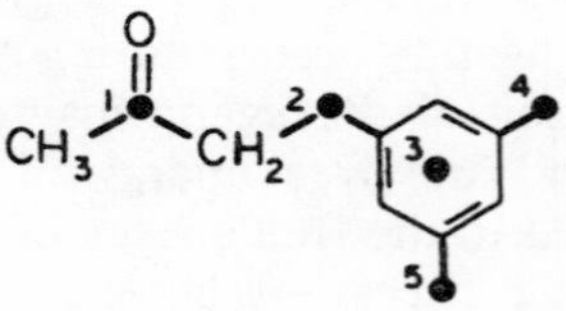

CH_3 — C(=O) — CH_2 — O — (benzene ring with positions 1, 2, 3, 4, 5)

Figure 5.1   The chymotrypsin inhibitor $m,m$-dimethylphenoxyacetone, decomposed into ligand points. The ligand points are assumed to lie at the heavy dots as indicated. Note that point 1 coincides with the carbonyl C, point 2 with the ether O, and points 4 and 5 with the $m$-methyls. Point 3 coincides with no atom but rather lies at the center of the benzene ring.

Table 5.1  Bound matrix, **B**, in Angstroms and ligand point types and numbering for *m,m*-dimethylphenoxyacetone.

| | type | $t_1$ | $t_2$ | $t_3$ | $t_4$ | $t_4$ |
| | no. | 1 | 2 | 3 | 4 | 5 |
|---|---|---|---|---|---|---|
| -CH$_2$COCH$_3$ | 1 | 0.0 | 2.6 | 5.5 | 7.8 | 7.8 |
| -O- | 2 | 2.6 | 0.0 | 3.0 | 5.2 | 5.2 |
| C$_6$H$_3$ | 3 | 3.5 | 3.0 | 0.0 | 3.1 | 3.1 |
| -CH$_3$ | 4 | 3.8 | 5.2 | 3.1 | 0.0 | 5.4 |
| -CH$_3$ | 5 | 3.8 | 5.2 | 3.1 | 5.4 | 0.0 |

could be correlated flexibility, as in the case of cyclohexane. If bond lengths and angles are taken to be fixed, it has only one degree of conformational freedom, so although both the C(1)-C(4) and the C(2)-C(5) distances have the same lower bound (achieved in the appropriate boat conformations) they cannot be attained simultaneously. Since the ligand points are matched with site points that have coordinates in three-dimensional space, other sorts of embedding constraints and correlations among the ligand points are automatically satisfied.

One similarly proposes a binding site in terms of a number of "site points" whose relative positions are specified primarily by $x,y,z$ coordinates and hence an embeddable fixed distance matrix. As in the case of the ligand points, the more detail that is required for the site, the more site points must be chosen. Whereas each ligand point represents some atom or group of atoms, the site points may be called either "empty" or "filled". An empty site point is a vacant place positioned where a ligand point may lie when binding takes place. For example, the simplest sort of hydrophobic pocket would consist of one such empty site point, and in binding, a phenyl group from the ligand molecule might coincide with that point. Similarly, hydrogen bonding or ion-pair sites may be represented by other empty site points. A filled site, however, indicates the position of some steric blocking group, and no ligand point may coincide with it during binding. In contrast to the ligands, the geometry of the site is represented by only a single distance matrix for the site points, instead of upper and lower bound matrices. Thus, the site is assumed to be rather rigid, with only some small variation allowed in each interpoint distance.

Representing both site and ligands as sets of points with conformations given by distance matrices has advantages in the calculation of binding which are crucial to the success of the method. Most importantly, a distance matrix is invariant under translation and rotation, so

the elaborate rigid body translation and rotation calculations involved
in the usual docking studies are totally avoided. Therefore, a possible
binding mode of some ligand amounts to simply a list of which ligand
points (not necessarily all) coincide with which empty site points (once
again not necessarily all). Filled site points must be avoided; at most,
one ligand point may occupy an empty site point; and the geometry of
the ligand points involved in binding must match that of their
corresponding site points. Details of the binding calculation are given
in the next section. The important point for now is that bringing
ligand and site together in a geometrically permissible and energetically
favorable way is treated as a relatively simple combinatorial (discrete)
problem, rather than a global optimization of energy with respect to 3
(continuous) degrees of translational freedom, 3 degrees of rotation,
plus whatever number of internal conformational degrees of freedom
that the ligand possesses.

The calculated free energy of binding is obtained in a simplified
all-or-nothing fashion by adding up the contribution from each contact
between a ligand point and a site point. The individual interaction
energy contributions are specified in a proposed energy table, where
each row corresponds to a type, $t_i$, of ligand point (e.g., methyl,
phenyl, carbonyl) and each column is for the site point (e.g., hydrophobic pocket, hydrogen bond acceptor). In general there will be
none, one, or several ligand points of a given type in one ligand; and
the same type of ligand point may appear in more than one molecule.
Each point–point interaction energy is taken to be the $\Delta G$ for the process: solvated ligand point + solvated site point $\rightarrow$ occupied site
point. Thus, solvation, enthalpy, and entropy are all included. As will
be explained in section 5.4, the interaction energies are determined
completely empirically by fitting the calculated $\Delta G_{bind}$ to the observed
free energies of binding. Frequently the experimental binding results
are given in the form of $I_{50}$, the millimolar concentration of an inhibitor required to produce 50% inhibition, but they can be converted, at
least approximately, to $\Delta G$ values of binding, assuming Michaelis-
Menten kinetics:

$$\Delta G_{bind} = +RT \ln \left[ \frac{K_m [I_{50}]}{K_m + [S]} \right] . \qquad (5.1)$$

Here $K_m$ is the Michaelis constant and $[S]$ is the substrate concentration used in the binding assay. The argument to the logarithm is the
equilibrium constant for the *dissociation* of enzyme and competitive
inhibitor, whereas $\Delta G_{bind}$ is the *association* free energy.

Before going into the details of carrying out a distance geometry
analysis of ligand binding as discussed in the following sections, we
should conclude this introduction with a brief overview of the general

procedure.

(i)     The structural information from each ligand is reduced to inter-
        point distance bounds and chirality relations, where the points
        represent the ligand molecule in sufficient detail for the purposes
        of the study.

(ii)    The structural variability of the set of ligands is analysed and
        compared to the observed binding energies in order to estimate
        how complicated the proposed binding site must be.

(iii)   The investigator proposes modes of binding for each of the
        ligands, and then the geometric consequences for the site struc-
        ture are deduced.

(iv)    Coordinates for the site points are located subject to the con-
        straints of the previous steps by the usual distance geometry
        algorithm.

(v)     The interaction energies are calculated as a least squares fit
        between observed free energies of binding and the calculated
        ones subject to the constraints that the proposed binding modes
        have the lowest energies of all geometrically allowed modes.

The heart of the whole approach is contained in step (v). The physi-
cally reasonable assumption is that the ligand will bind at the site in
whatever orientation and conformation that minimizes its free energy
of binding. This implies that one must examine all possible binding
modes and select the geometrically allowed one of lowest energy.
Exactly how this is done will be discussed in the next section.

### 5.2 Geometric Constraints on Binding

Before going into detail about calculating the optimal binding
mode, additional explanation about the representation of the ligand
molecules is necessary. It is relatively easy to obtain coordinates for
all atoms of a ligand, including hydrogens. With luck, a crystal structure is available for the compound itself or at least a closely related
analogue. Reasonable modifications, if necessary, can be done using
standard bond lengths and angles. Then all possible combinations of
dihedral angles for all rotatable bonds are tried, generating the new
coordinates from the original set. Sterically disallowed conformations
are rejected in a careful fashion, so as to "prune" the tree of all possibilities and shorten the search. It is this stage particularly that requires
the hydrogens, since then one needs only to check the calculated
interatomic distances against Van der Waals radii of individual atoms,
rather than those of atom groups. For every allowed conformation, all
interatomic distances are examined to find the global minimum and
maximum of each. Even with a carefully constructed tree search, the
time required usually is proportional to $p^q$, where $p$ is the number of

values each of the $q$ dihedral angles may take on. Even for $p=3$ and $q=12$, the search is quite time consuming.

Once the upper and lower distance bounds have been derived, the chirality can be found. Whenever a quartet of atoms have mutual distances that are relatively invariant, say $u_{ij}-l_{ij}<0.1$, and are not coplanar, then the four atoms constitute a *chiral center*. Unlike eq. 3.2, we are no longer interested in the magnitude of the volume covered by the four points, but only its sign. Let $\mathbf{w}_{ij}$ be the normalized vector from point $i$ to $j$:

$$\mathbf{w}_{ij} = (\mathbf{c}_j - \mathbf{c}_i)/\|\mathbf{c}_j - \mathbf{c}_i\|$$

where $\mathbf{c}_i$ is the position of point $i$ in the original set of coordinates. Then

$$g_{ch} = \mathbf{w}_{ij} \cdot (\mathbf{w}_{ik} \times \mathbf{w}_{il})$$

gives the chirality of the ordered quartet of atoms, $i, j, k, l$. If $|g_{ch}|<0.15$, the quartet is essentially coplanar, and hence achiral; otherwise, the sign of $g_{ch}$ gives the handedness of the quartet. An odd permutation of the ordered quartet will reverse the sign of $g_{ch}$, but an even permutation leaves it unchanged. As usual in these calculations, each atom has its unique label, so that even methylene groups are "chiral" in spite of the two indistinguishable hydrogen substituents. Aside from this minor problem, the definition is extremely general and correct, encompassing ordinary asymmetric carbon atoms, spiranes, and even chirality due to steric hindrance.

In sum, a ligand molecule is described as an interatomic distance bound matrix, $\mathbf{B}$, a list of atom types (reflecting functional properties, pK, etc.) and a list of chiral quartets, each entry of which consists of the ordered list of the four atoms involved and the sign of $g_{ch}$. A full one point per atom representation has been necessary up to this point in order to correctly treat rotations about single bonds and steric hindrance. However the later stages of the calculation will be greatly speeded by using fewer points to describe the molecule. Removing unnecessary atoms, such as hydrogens, is simply a matter of deleting the corresponding rows and columns from $\mathbf{B}$ and removing any chiral quartets involving them. Sometimes this also requires alterations in the atom types in order to still reflect functional groups properly. For example, suppose originally an aldehyde group was given as atoms of types "hydrogen", "trigonal carbon", and "double bonded oxygen", while a carboxylate ion was given as "trigonal carbon" and two "charged oxygens". Then if the molecule representation is simplified by removing all hydrogens and oxygens, both groups would consist of the same single trigonal carbon.

Now suppose we are given a ligand in its point representation, coordinates for the hypothesized site points, and a hypothesized table of interaction energies. The task is to find the geometrically allowed binding mode (pairing up of ligand and site points) with the lowest calculated binding free energy. This is quite close to the *bipartite matching* problem from graph theory, for which there exists an efficient (polynomial time) algorithm (Lawler, 1976). In that situation, there are two non-intersecting sets of points (our ligand and site points) and given weights on edges connecting a member of one set with a member of the other (our interaction energy table). The bipartite matching algorithm finds the set of edges with the optimal sum of weights such that every point is involved at most one time (which is equivalent to our restriction that a ligand point can contact at most one site point, and a site point can be occupied by at most one ligand point). The only problem is that the geometric restrictions on which combinations of contacts are allowed cannot be easily included in the bipartite matching algorithm. Therefore, I have employed a tree search of all possible binding modes, as follows:

(1) Generate successively all possible combinations of contacts of ligand points with each site point, including the possibility that some of the site points may have no ligand points in contact with them.

(2) Reject any contact combination which has one ligand point in contact with two site points (or of course one site point in contact with two ligand points), although several unused site or ligand points are allowed.

(3) Reject any contact combination which includes contact with an unfavorable (i.e., positive) interaction energy.

(4) Check that for each pair of used site points, $i$, and $j$, the distance between them, $d_{ij}$, is in the range of distances allowed to the corresponding ligand points, I and J, with which they are respectively in contact. That is, $u_{IJ}+\delta \geqslant d_{ij}$ and $l_{IJ}-\delta \leqslant d_{ij}$ both hold, where $u_{IJ}$ and $l_{IJ}$ are the upper and lower bounds, respectively, on the distance between ligand points $I$ and $J$; and $I$ is in contact with site point $i$ and $J$ with site point $j$. The parameter $\delta$ represents the allowed flexibility in the site.

(5) Having now found a contact combination which is geometrically allowed and contains no unfavorable energy interactions, it is now necessary to determine if some (possibly energetically unfavorable) contacts follow as the necessary consequences of those already chosen. These are referred to as "forced contacts" and are classified as being the consequence of certain combinations of three contacts, two contacts, or perhaps even of one contact. Each case is considered, in turn, in this order.

(6) A triple-point forced contact occurs whenever there is an unused ligand point whose position is fixed in space by having rather

invariant distances to three other ligand points which are in contact with some three site points, *and* there is a fourth site point having distances to the other three site points which match the three invariant distances of the ligand points. For example, suppose we were fitting neopentane into a tetrahedral 4-point site, one point for each methyl group. If a proposed contact combination paired methyl groups 1, 2, and 3 with their corresponding sites 1, 2, and 3, then necessarily methyl no. 4 would have to lie on site no. 4. The invariant distance proviso requires that the upper and lower bounds on the interligand points differ by no more than an arbitrary upper limit, taken to be 1. That way the fourth ligand point is accurately triangulated in space and has no way of avoiding a correspondingly placed site point. Now there are four mutually rigid ligand points in contact with four site points, so if the four ligand points constitute one of the known chiral quartets, then their handedness must match that of the four site points. Care must be taken to preserve the order of points in such comparisons. In addition, any new forced contact is checked for geometric compatibility with the existing contacts. If the test of step 4 is failed, then the whole contact combination is rejected.

(7)   After all triple-point forced contacts have been deduced, a weaker sort of double-point forced contact is considered. When an unused ligand point has two invariant distances to some two used ligand points, the position of the unused one is restricted to lie on a circle in space. If there are four or more site points which also lie on that circle (as indicated by their having the same corresponding distances to the two site points involved in contacts with the two used ligand reference points), then the unused point in question must be in contact with one of these four site points. Take the new contact to be with the energetically most favorable unused site point of the four. Of course, the choice of 4 as the number of site points to completely occupy a circle in space is rather arbitrary, and increasing the number would amount to making the search more detailed. The main use of this sort of forced contact is to require a part of the ligand that can rotate to either find an energetically favorable orientation in the site or be excluded altogether in the case that all four site points are repulsive.

(8)   The last sort of forced contact to be considered is the single-point variety, where making one contact constrains an unused ligand point with invariant distance to the used one, to lie on a spherical shell in space. If there are six site points that also lie on this shell, then the unused ligand point is taken to be in contact with the most energetically favorable unused one. This is certainly the situation in a concave site "pocket", and once again the

choice of six is arbitrary, except that it should clearly be greater than the number of site points necessary to force a double-point contact.

(9)   As indicated above, triple-point forced contacts are the most specific and are tried first until no more can be made. Only then are double-point contacts attempted. If one is formed, then perhaps triple-point contacts can again be deduced, so that must be tried again exhaustively. Only when no more triple- or double-point contacts can be formed, are the single-point forced contacts tried. Once again, success results in trying triple-points again. When at last no contacts of any variety can be forced, the proposed contact combination is considered to be complete.

(10)   If all four points of a chiral quartet are involved in the (possibly revised) contact combination, then the chirality of the ligand points and the site points must match. This condition is tested for all chiral centers.

(11)   Finally, if the combination has passed all the above tests, its energy is evaluated simply by summing the energetic contributions of each contact. The contribution is taken to be the given interaction energy table entry for the corresponding ligand point type and site point. Unused ligand or site points contribute zero to the sum. the calculated mode of binding is the contact combination that gives the minimal calculated binding energy.

## 5.3 Framing the Hypothesis

It is not always easy to tell initially just how complicated a site must be in order to explain the observed binding energies, much less the site geometry, so some preliminary systematic calculations are required. Assume for the moment that each ligand binds to the site in the same orientation as all the others, with common features in contact with always the same site points. The picture is that of a site having some number of points dedicated to the common structural feature of the set of ligands and other surrounding site points to accommodate any occurrence of a substituent at the various points of attachment. Of course this approach is not new, being essentially equivalent to the linear-free-energy or Free-Wilson method (Martin, 1978), but here we will concentrate on the geometric interpretation of the results of the analysis, outlined as follows.

*Decomposition algorithm.* Given the chemical structures of all the ligands, coded as ligand points of various types with given upper and lower bound distances among the points and any chiral centers, we wish to find the common structural feature, if any, and a small set of substituent groups that describe the differences. The substituent groups will be characterized by the number of ligand points involved, their types, their relative positions as specified by upper and lower distance bounds, and by their locations relative to the base group as specified by upper and lower distance bounds. Therefore it is necessary that the base consist of at least three points (and preferably four) which are not all colinear (and preferably four which are not all coplanar) in order to correctly fix the locations of the substituents. Of course, it is possible that there may be no such base group common to *all* the ligands, in which case it would be desirable to subdivide the original set of ligands into more than one subset, each of which would have a common base. Once the base group has been determined, there are still a variety of ways to choose the substituents. For example, if at a given position in a series of analogs, there is always either a -H or a $-CH_2CH_3$, it is reasonable to deduce two substituents, -H and $-CH_2CH_3$, rather than three groups, -H, $-CH_2-$, and $-CH_3$. However, if an isopropyl group also occurs in that position, the latter set of substituents would be preferable, since the isopropyl group could be described as a methyne (taken to be the same type as a methylene) and two methyls. The policy I have taken is to choose the smallest number of substituents necessary to describe the entire set of ligands, even if that means having to represent any one ligand as a greater number of parts. The algorithm proceeds from these considerations:

1.  *Determine base group.* Of all the ways to pair some of the points in the first ligand with an equal number of points in the second, there is an optimal, not necessarily unique, matching that involves the greatest number of points such that the corresponding points in the two ligands have the same type and such that the range of allowed distances between two points in the first ligand overlaps the corresponding range in the second ligand, for all pairs of points used in the matching in the first ligand. Let the "intersection" of two ligands be the collection of points in the optimal matching, their types, and the set of upper and lower bounds among them, always choosing for a given pair of points in the intersection the greater lower bound between the two ligands and the lesser upper bound. The matching is carried out by exhaustive enumeration of all the viable possibilities in a tree search that excludes classes of disallowed matches whenever possible. The computational effort required for the search increases rapidly when there are many points of the same type which must be distinguished on geometric grounds only. Clearly the ordering of the points within a single ligand has no effect on the matching process, and the intersection always has no more points than the smaller of the two ligands. To find the base group of a set of ligands, first take the intersection of the first and second ligands; then take the intersection of the result with the third ligand; and so on until all ligands have been considered. If at any time in

the process, the number of points in the intersection should drop below 3, the latest ligand is declared to have no recognizable common structural feature with the others and is excluded from further consideration. The base group is the resultant intersection. It is certainly possible that the results depend on the sequence of the ligands when some of the ligands must be excluded, and the algorithm does not attempt to overcome this shortcoming.

2.    *Determine substituent groups.* Having found a subset of ligands with a common base in the previous stage of the calculation, we now remove the base points from each ligand. The definition of intersection required in this step is the same as before except that in addition, the ranges of distances from ligand point to the constant base must also overlap, and the overlap distance ranges to the base are included in the description of the intersection. For the points remaining in the first ligand and those left in the second ligand, find the intersection according to the above definition. Intersect the result with the remaining points of the third ligand and so on until no points remain in the intersection. Then the outcome of the previous intersection is a substituent group, and every occurrence of it in the set of ligands is removed from further consideration. Beginning the process again with the first ligand that still has some points left, determine the next substituent group, iterating until every point of every ligand has been accounted for.

3.    *Termination.* The initial set of ligands has now been completely broken down into a single common base group and a collection of substituents occurring one or more times each in some but not all the ligands. However there may be other ligands that were excluded in step 1 on the grounds of having no common base with the others. We simply repeat steps 1 and 2 with these alone, independently of the results obtained so far. If any ligands still are excluded, iterate until all have been taken care of. Clearly the worst case is that every ligand stands alone in its separate class, having no common structure with any of the others.

As an illustration of the above procedure, consider a set of 68 quinazoline inhibitors of dihydrofolate reductase (Crippen, 1980). The base group turned out to be the quinazoline ring system plus a phenyl ring (or half a naphthyl group) that all the analogues happened to have attached to either the 5- or 6-position on the quinazoline. Due to flexibility of the linkage between quinazoline and phenyl, attachment to either position gave an overlapping range of distances between the two ring systems. For technical reasons, 2 analogues were found to not quite share the common base group, and were excluded from the subsequent substituent analysis. The decomposition algorithm went on to find 33 different substituent groups, of which only about 14 are geometrically distinct. Most substituents were groups represented only by a single point, and corresponded to functional groups that a chemist would tend to pick out, such as the 2-NH$_2$, 5-S-, etc. However some are rather complex, as in the case of a CONHCH group.

Now if the constant binding mode assumption is correct, a least squares fit of the observed $\Delta G_{bind}$ values to the sum of the contributions of each component group in a ligand for all ligands should have a low residual error. In the quinazoline example, however, the 68 observations could not be matched within ±1 kcal even using 34 parameters! Given the model used, this is an objective determination that the constant binding mode assumption is in error at the 1 kcal level, since any other choice of interaction energies would give an even worse fit to the data. However it is still not clear *which* ligands violate the constant mode assumption. The nonconforming ligands can be identified by removing the worst offender from the data set, repeating the fit, once again removing the worst ligand, and so on until the observed $\Delta G$ values for all the remaining inhibitors can be fit to the desired accuracy.

Suppose for a moment that the constant binding mode analysis *does work*. Then proposing a distance geometry site and interaction energies is now quite easy. One site point is chosen for each point in the base group and for each point in each geometrically distinct substituent group. The relative positions of the base group points are known as are the positions of each substituent relative to the base, so calculating all the coordinates is a straightforward procedure according to the standard distance geometry algorithm. The foregoing least squares calculation has already produced the interaction energies required between site points and the contacting ligand points. Let all other interaction energies be positive to discourage alternate binding modes.

On the other hand, suppose that as in the case with the quinazolines, the fixed binding mode analysis fails, the most striking exceptions to the constant binding mode assumption have been identified, and we must set out on the *variable mode analysis*. At this point the distance geometry approach to QSAR becomes a vehicle for formulating and testing the investigator's relatively subjective hypotheses. Usually the minimum number of site points required would be three for the base group and one for each geometrically distinct substituent group. More are required if very different binding modes are imagined for some of the ligands, or when binding pockets are outlined with filled site points. Perhaps one analogue binds poorly to the actual protein because it cannot reach far enough to form a hydrogen bond with some amino acid residue at the optimal distance, and must settle for a stretched one instead. In spite of the all-or-nothing nature of the calculated binding energies, this effect can be modeled by having two site points represent the residue: one at the optimal hydrogen bond length position with a corresponding very favorable binding energy to appropriate ligand points, and the other at the stretched hydrogen bond length position with a corresponding less favorable interaction energy. In the case of the quinazoline analogues binding to dihydrofolate reductase, it was hypothesized for various reasons that the entire quinazoline ring and its immediate substituents must lie at a different tilt in the site, depending on whether it was 2,4-diamino substituted. Thus all ligand points rigidly attached to the ring system and not on the proposed axis of rotation required two sets of site points. The best procedure is to draw up a list of site points and their assumed roles in binding: what ligand points contact each under what circumstances. Then it is easy to set down in more detail the *desired binding modes* of each ligand. Once that is done, a great deal has implicitly been said about the geometry of the site, and it is a straightforward matter to devise an algorithm to deduce the upper and lower bounds on the distances among the site points. It is important to remember that the calculation of optimal binding mode given a ligand and a full site description allows the site the same small flexibility, $\delta$, in each interpoint distance. Thus a contact is allowed between ligand point $La$ and site point $Sa$, when for every other contact between ligand point $Lb$ and site point $Sb$, both $d_{Sa,Sb}+\delta \geqslant l_{La,Lb}$ and $d_{Sa,Sb}-\delta \leqslant u_{La,Lb}$. Hence, the algorithm for deducing upper and lower bounds on the distances among the site points proceeds as follows:

1.  Initially let the upper bounds be some large value and the lower bounds be zero.

2.  For each ligand in turn, consider every pair of contacts $La-Sa$ and $Lb-Sb$. If the site point upper bound, $u_{S:Sa,Sb}$, is greater than the corresponding ligand point upper bound, $u_{L:La,Lb}+\delta$, then set the site point upper bound to the latter value.

3.  Similarly, whenever $l_{S:Sa,Sb} < l_{L:La,Lb}-\delta$, the site point lower bound is raised to the latter value.

4.  If the desired binding modes are unfortunately chosen, at some point in the process, a site point lower bound will exceed the corresponding upper bound. In such a case, it is clear that the desired binding mode of the ligand under present consideration is geometrically incompatible with that of some preceding ligand. A more detailed analysis enables one to pinpoint at least two conflicting ligand binding modes and the relevant desired contacts. Even if there is no conflict, inconsistencies may appear later if the $\delta$ used in this calculation is not slightly smaller than the $\delta$ for the subsequent optimal binding mode calculations. The reason is that the distance geometry algorithm for producing

coordinates of site points from the bounds deduced here tends to result in some distances that are equal to or very slightly beyond the specified bounds. In such cases, and equal δ value in the optimal binding calculation may yield very slight geometric violations when attempting the desired binding mode, whereas a somewhat larger δ permits the desired mode.

5.   Whenever all four points of a chiral center for some ligand are to bind to site points in the desired binding mode, then the chirality of those four site points must match that of the chiral center. Thus chiral constraints are introduced as well as distance bounds.

By the distance geometry algorithm outlined in section 3.4, the constraints deduced above from the desired binding modes are used to produce a sampling of possible site point coordinate sets. It is possible that due to errors in proposing the binding modes, there is no arrangement of the site points in space which will meet the deduced geometric constraints. Fortunately, one can usually trace back to the source of such inconsistencies. It is important to be confronted with the range of site configurations allowed by the desired binding modes. In the quinazoline example, the angle of tilt between the two hypothesized modes was not determined by the desired modes, so that site structures were produced ranging from 20° to 180° angles. Presumably the further analysis would proceed equally well with any of the site configurations.

### 5.4 Fitting the Data

So far, the variable binding mode analysis has concentrated entirely on geometry. Now holding the site geometry fixed, we find empirical interaction energies to fit the observed binding data. The objective function, $F$, must be minimized, where

$$F \equiv \sum_{ligands} (\Delta G_{obsd} - \sum_{\substack{desired \\ contacts}} \epsilon_{ij})^2 \, ,$$

and the $\epsilon_{ij}$ are the interaction energies to be determined. Since the calculated binding free energy is simply the linear sum of the interaction energies in the binding mode, the optimization is a linear least squares problem. However in addition we must require that the desired binding mode has a more favorable (lower) calculated energy than that of any other. This produces linear inequality side constraints having the form

$$\sum_{\substack{desired \\ mode}} \epsilon_{ij} < \sum_{\substack{other \\ mode}} \epsilon_{ij} \, ,$$

but since there are a large number of undesirable yet geometrically allowed binding modes for each of the several ligands, there are very many inequalities. At least the form of the linear least squares optimization subject to linear inequality constraints is exactly a *quadratic programming* task, which can be reliably solved by standard methods such as the Wolfe algorithm (Kuester and Mize, 1973). As implemented, the quadratic programming step does not use a starting guess for the energy parameters, so the values found in the constant binding mode analysis are not needed here. Since there are very many geometrically allowed alternate binding modes, and only a few of these contribute significant inequalities to the problem, the approach is to first optimize without constraints, note the first ligand that found an energetically optimal binding mode other than the desired one given the energy parameters, add the corresponding inequality to the problem to eliminate the unwanted binding mode, reoptimize with current constraints, and so on. The iteration ends when at last the energy parameters that give an optimal fit to the observed binding free energies for the desired binding modes also give a higher (worse) calculated binding energy for any other mode for all ligands.

With this method of refining the energies, one can give a good accounting of the number of degrees of freedom. In the quinazoline case, 20 parameters were adjustable, i.e., were involved in desired binding modes. At the final optimum, 1 of these was set to zero by the quadratic programming algorithm; that is, it was "not used". By the time the final iteration was reached, 27 inequalities had been generated due to alternate better binding modes found along the way by some 20 of the ligands. Of these inequalities, only 11 were still "active" at the solution, which means that they were restraining the optimal solution from moving into their disallowed zones. Hence there were $20-1-11=8$ adjustable energy parameters.

One might well ask if there is enough information in a set of binding data such as these 68 quinazolines to determine all the geometric and energetic parameters describing the site. After all, other QSAR methods produce only a few parameters to define their fit (cf. Hansch et al. (1977) using the same data set employ only 6 empirical parameters). Mathematical information theory (Ash, 1965) provides at least a rough answer by the following "order of magnitude" argument. The site, its geometry and interaction energies, can be thought of as a source of a lengthy message. Each ligand, along with its observed free energy of binding and its (flexible) structure, corresponds to a signal sent to us over a noisy channel, where the noise is the experimental error in the binding energies and the conformational variability of the ligand. The question then becomes: how many signals must be sent at least in order to deduce the original message, namely the site? The information theory measure of uncertainty is

$$H \equiv \sum_i p_i \log_2 p_i \quad ,$$

where the $p_i$ are the probabilities of receiving the $i$-th type of signal, assuming the different possibilities are mutually independent. If some events are dependent, $H$ will be less. $H$ attains its maximum value, $\log_2 n$, when each of the $n$ possibilities are equally probable. We first crudely estimate the maximum possible uncertainty in the proposed site in the quinazoline study, and then from an equally rough estimate of the information conveyed by each ligand, we can show how many ligands must be included in the data set to reduce the site uncertainty to zero. In the geometry of the site, for the proposed 11-point quinazoline binding site (Crippen, 1980), there are $11 \times 3 - 6 = 27$ possibly mutually independent coordinates to be determined to an accuracy of $\pm \delta = \pm 0.5$, the site flexibility parameter in that study. The coordinates of the proposed site ran over at most a 10 range each, so with the given accuracy, there are 10 distinguishable possibilities for each coordinate. Therefore there are at most $H_{S:coord} = 27 \log_2 10 = 89.69$ bits of uncertainty in the site coordinates. Similarly we suppose the 20 possibly independent energy parameters need to be determined within $\pm 0.1$ kcal out of a range of 3.4 kcal, the difference between the largest and smallest interaction energies in the final optimized set. Hence there are 17 distinguishable energy levels for each parameter and $H_{S:energy} = 20 \log_2 17 = 81.75$ bits. The total maximum estimated uncertainty in the site, $H_S = H_{S:coord} + H_{S:energy} = 171.4$ bits. The optimistic view is that the uncertainties in the data set, which measure the maximum number of bits of information obtainable from it, can be fully used to determine the large uncertainty in the site. The data set as a whole (considering now only the 66 ligands having a common base group under the decomposition algorithm) contains structural variability in the form of the different substituents. The base group does not count because it is found in all the 66 ligands, and is therefore completely predictable. Of the 33 substituent groups found by the decomposition algorithm, 9 were *geometrically* identical to some of the others, and therefore do not contribute to the variability. If the regions of space for each substituent (as determined by their upper and lower distance bounds to the base group) are all equal in volume and do not overlap (i.e., the most favorable case) then the total data set geometric uncertainty is $24 \log_2 24 = 110.04$ bits. If all this variability can be fully applied to deducing the site, then there remains $171.4 - 110.0 = 61.4$ bits to be determined. Now the range in observed binding free energies was only about 8 kcal, and since the estimated error may by $\pm 1$ kcal, there are only 4 distinguishable energy levels. The distribution of $\Delta G_{obsd}$ values is fairly even over the data set and yields $H_{L:energy} = 1.971$ bits as the average uncertainty in energy per ligand. It is interesting to note how small this value is, which is due to the small range of values and large experimental error. In addition, each ligand also has an uncertainty due to its composition in terms of substituents. The geometry of each substituent has already been accounted for. Assuming the distribution of substituents throughout the set of ligands is mutually independent, the known frequencies of occurrences of each substituent in the data set yields $H_{L:comp}=4.4$ bits as the average uncertainty per ligand in structure. Note that more information is potentially conveyed in the chemical structure of ligands than in their binding energies. The most information that could be extracted per ligand then is $H_{L:energy}+H_{L:comp}=6.4$ bits, assuming that composition and binding energy *appear* to be independently distributed. Our very rough estimate for the minimum number of ligands we need to observe is $61.4/6.4=10$ ligands. Thus we can say that deducing a site of the complexity given is at least conceivable.

# BIBLIOGRAPHY

A. C. Aitken (1956), "*Determinants and Matrices*", 9th edn., Interscience Publishers, New York, pp. 72-73.

R. Ash (1965), "*Information Theory*", Wiley Interscience, New York.

L. M. Blumenthal (1970), "*Theory and Applications of Distance Geometry*", 2nd edn., Chelsea, New York, pp. 98-105.

W. Braun, C. Boesch, L. R. Brown, N. Go, and K. Wuethrich (1980), submitted.

Cahn, Ingold, and Prelog (1966), *Angew. Chem. Int. Ed. Engl., 5*, 385.

M. L. Connolly (1977), private communication.

G. M. Crippen (1977), *J. Comp. Phys., 24*, 96-107.

G. M. Crippen (1978), *J. Comp. Phys., 26*, 449-452.

G. M. Crippen (1979a), *Int. J. Pept. Prot. Res., 13*, 320-326.

G. M. Crippen (1979b), *J. Med. Chem., 22*, 988-997.

G. M. Crippen (1980), *J. Med. Chem., 23*, 599-606.

G. M. Crippen and T. F. Havel (1978), *Acta Cryst. A, 34*, 282-284.

G. M. Crippen, N. J. Oppenheimer, and M. L. Connolly (1980), *Int. J. Pept. Prot. Res.*, in press.

D. K. Faddeev and V. N. Faddeeva (1963), "*Computational Methods of Linear Algebra*", Freeman, San Francisco, pp. 307, 328-330.

R. Fletcher and C. M. Reeves (1964), *Comput. J., 7*, 149.

A. O. Griewank, B. R. Markey, and D. J. Evans (1979), *J. Chem. Phys., 71*, 3449-3454.

M. Hall, Jr. (1967) "*Combinatorial Theory*", Blaisdell Publishing Co., Waltham, Mass., p. 8.

J. M. Hammersley and D. C. Handscomb (1964), "*Monte Carlo Methods*", Methuen, London, pp. 25-42.

C. Hansch, J. Y. Fukunaga, P. Y. C. Jow, and J. B. Hynes (1977), *J. Med. Chem., 20*, 96.

T. F. Havel, G. M. Crippen, and I. D. Kuntz (1979), *Biopolym., 18*, 73-81.

J. L. Kuester and J. H. Mize (1973), "*Optimization Techniques with Fortran*", McGraw-Hill, New York, pp. 106-119.

I. D. Kuntz and G. M. Crippen (1980), *Biophys. J.*, in press.

E. L. Lawler (1976), "*Combinatorial Optimization: Networks and Matroids*", Holt, Rinehart & Winston, New York, ch. 5.

Y. C. Martin (1978), "*Quantitative Drug Design*", v. 8 in Medicinal Research Series, G. L. Grunewald, ed., Macel Dekker Inc., New York.

# INDEX